Annals of the ICRP

ICRP PUBLICATION 115

Lung Cancer Risk from Radon and Progeny and Statement on Radon

Editor
C.H. CLEMENT

Authors on behalf of ICRP
M. Tirmarche, J.D. Harrison, D. Laurier, F. Paquet,
E. Blanchardon, J.W. Marsh

PUBLISHED FOR

The International Commission on Radiological Protection

by

Los Angeles | London | New Delhi
Singapore | Washington DC

Please cite this issue as 'ICRP, 2010. Lung Cancer Risk from Radon
and Progeny and Statement on Radon.
ICRP Publication 115, Ann. ICRP 40(1).'

CONTENTS

SAGE

ICRP Publication 115

Annals of the ICRP

Guest Editorial

RADON

Radon is a recognised cause of lung cancer. Radium-222, a gas, is a member of the uranium-238 decay chain. As its immediate precursor, radium-226, is ubiquitous in the Earth's crust, radon is present in all buildings and underground locations. Radon is a significant source of radiation exposure to the general public and, in some situations, can be a main source of exposure at work. Levels of exposure can, however, vary hugely depending on the local geology, the type of building, its ventilation, and the behaviour of the occupants.

The importance of radon as a source of exposure, together with the fact that radon levels in buildings can, at least in principle, be controlled, has prompted the Commission to issue recommendations for protection against radon. These recommendations were provided in 1993 as *Publication 65* (ICRP, 1993), and are framed in the context of the Commission's system of radiological protection (ICRP, 1993). The Commission's protection policy for radon is based on setting an annual effective dose level of around 10 mSv for radon, where action would almost certainly be warranted to reduce exposure. This dose is converted into practical action levels set in terms of Bq/m^3 using a dose conversion convention. The Commission's policy forms the basis for protection against radon worldwide. Recently, the Commission issued the 2007 Recommendations (ICRP, 2007), which formally replace the 1990 Recommendations (ICRP, 1991). The 2007 Recommendations distinguish between planned and existing exposure situations of radiation exposure. Most radon exposures are existing situations because the source of exposure is present when a decision on control has to be taken. Protection against radon is achieved by application of reference levels and optimisation.

An understanding of the health risk from radon exposure is fundamental to setting the reference levels. This Task Group report on lung cancer risk from radon provides current information on health risks from radon by reviewing recent epidemiological studies on residential and occupational exposures. An important conclusion is that the detriment-adjusted nominal risk coefficient for exposure to radon should now be taken to be around twice that assumed previously (ICRP, 1993). Furthermore, radon appears to act in a more multiplicative than additive manner on the underlying rates of lung cancer of the exposed population. Thus, for the same radon exposure, the risk of lung cancer from radon for smokers is substantially greater than that for non-smokers. Although comparisons are complex, lifetime risk estimates from residential exposure are consistent, as far as one can tell, with those estimated for underground miners at low levels of exposure, which adds strength to the overall conclusions.

The interaction with smoking is challenging. It raises a question about whether ICRP should set its protection standards for radon for smokers, non-smokers, or, as currently, a mixture of both. It is important to remember that the purpose of the system of protection is to control sources of exposure and exposures, not radiation risks to specific individuals. The system is also intended for worldwide application. Its primary dosimetric quantity, effective dose, estimates a 'dose' to a reference individual including characteristics averaged across ages and both sexes. The tissue-weighting factors, which represent the relative radiosensitivities of tissues, are judgements based on transferring radiation risks across populations with sometimes widely differing baseline cancer rates, using a mixture of multiplicative and additive risk projection models. Effective dose is a defined radiation protection quantity which can change over time as new judgements are made with regard to tissue- and radiation-weighting factors. Effective doses provide neither the best estimates of doses nor risks to individuals. However, the Commission continues to believe that its system of protection, including the effective dose quantity, remains the most appropriate approach for protection against exposures and sources of exposure. Further, attempting to differentiate within the system of protection between individuals on the basis of lifestyle factors not directly related to radiation exposure would result in unjustified complexity without improved protection, and a system that was unnecessarily burdensome and could be discriminatory.

Also published in this volume is the Commission's Statement on Radon from its 2009 meeting in Porto, Portugal. The Statement takes account of the important findings from this Task Group report, revising downwards the upper reference level for radon in dwellings in line with the change in the nominal risk coefficient for radon. On a similar basis, the Commission also revises downwards the reference level for workplaces, recommending a single value of 1000 Bq/m^3 which serves as an entry point for applying occupational radiological protection requirements. Importantly, the Commission also announces its intention to replace the current dose conversion convention with a dosimetric approach, bringing radon into line with all other internal emitters. The dosimetric approach considers a range of parameters relevant to doses from radon, the values for which may change depending on the circumstances of exposure. Thus, any given concentration of radon may result in different doses depending on the circumstances. The Commission will, therefore, be reconsidering its policy for protection against radon when the dosimetric approach is finalised, to ensure it is coherent and proportionate.

Lung cancer from radon exposure has clearly occurred in uranium and other underground miners for centuries, although radon was only recognised as the culprit in the last century. This report on lung cancer risk from radon will contribute to protection against radon into the 21st Century. The Commission is currently preparing practical advice on how to implement its new recommendations for protection against radon in dwellings and the workplace.

JOHN COOPER
ICRP MAIN COMMISSION MEMBER

References

ICRP, 1991. The 1990 Recommendations of the International Commission on Radiological Protection. ICRP Publication 60. Ann. ICRP 21 (1–3).

ICRP, 1993. Protection against radon-222 at home and at work. ICRP Publication 65. Ann. ICRP 23 (2).

ICRP, 2007. The 2007 Recommendations of the International Commission on Radiological Protection. ICRP Publication 103. Ann. ICRP 37 (2–4).

Lung Cancer Risk from Radon and Progeny

ICRP PUBLICATION 115, PART 1

ICRP Publication 115

Lung Cancer Risk from Radon and Progeny

ICRP PUBLICATION 115, PART 1

Approved by the Commission in April 2011

Abstract–Recent epidemiological studies of the association between lung cancer and exposure to radon and its decay products are reviewed. Particular emphasis is given to pooled case–control studies of residential exposures, and to cohorts of underground miners exposed to relatively low levels of radon. The residential and miner epidemiological studies provide consistent estimates of the risk of lung cancer, with significant associations observed at average annual concentrations of approximately $200\ \mathrm{Bq/m^3}$ and cumulative occupational levels of approximately 50 working level months (WLM), respectively. Based on recent results from combined analyses of epidemiological studies of miners, a lifetime excess absolute risk of 5×10^{-4} per WLM $[14 \times 10^{-5}$ per $(\mathrm{mJh/m^3})]$ should now be used as the nominal probability coefficient for radon- and radon-progeny-induced lung cancer, replacing the previous *Publication 65* (ICRP, 1993) value of 2.8×10^{-4} per WLM $[8 \times 10^{-5}$ per $(\mathrm{mJh/m^3})]$. Current knowledge of radon-associated risks for organs other than the lungs does not justify the selection of a detriment coefficient different from the fatality coefficient for radon-induced lung cancer.

Publication 65 (ICRP, 2003) recommended that doses from radon and its progeny should be calculated using a dose conversion convention based on epidemiological data. It is now concluded that radon and its progeny should be treated in the same way as other radionuclides within the ICRP system of protection; that is, doses from radon and its progeny should be calculated using ICRP biokinetic and dosimetric models. ICRP will provide dose coefficients per unit exposure to radon and its progeny for different reference conditions of domestic and occupational exposure, with specified equilibrium factors and aerosol characteristics.
© 2011 ICRP Published by Elsevier Ltd.

Keywords: Radon; Lung cancer; Radiological protection; Risk

AUTHORS ON BEHALF OF ICRP

M. TIRMARCHE, J.D. HARRISON, D. LAURIER,
F. PAQUET, E. BLANCHARDON, J.W. MARSH

Reference

ICRP, 1993. Protection against radon-222 at home and at work. ICRP Publication 65. Ann. ICRP 23(2).

PREFACE

A Task Group of ICRP Committee 1 was established by the Commission in 2005 to investigate and report on risks from alpha-emitting radionuclides. In 2007, after a period of initial data gathering, this Task Group was asked to focus initially on the risks from radon and its progeny. Membership of the Task Group includes members of ICRP Committees 1, 2, and 4.

The present report reviews epidemiological studies of lung cancer associated with the inhalation of radon and its progeny in homes and underground mines.

The work of the Task Group continues with investigation of risks from other alpha-emitting radionuclides.

The membership of the Task Group was as follows:

M. Tirmarche (Chairperson)	J.D. Harrison	F. Paquet
M. Blettner	D. Laurier	N. Shilnikova
E. Blanchardon	J.F. Lecomte	M. Sokolnikov
E. Ellis	J.W. Marsh	

The corresponding members of the Task Group were:

B. Grosche	J. Lubin	C.R. Muirhead

The consultants providing advice were:

F. Bocchichio	L. Tomášek	D. Chambers

The reviewers of report drafts were:

J. Boice	D. Chambers	J. Lochard

The membership of Committee 1 from 2005 to 2009 was:

J. Preston (Chairman)	C.R. Muirhead	M. Tirmarche
A. Akleyev	R Ullrich	P.-K. Zhou
M. Blettner	D.L. Preston	
R. Chakraborty	W. Rühm	
J. Hendry	R.E. Shore	
W.F. Morgan	F.A. Stewart	

The membership of Committee 2 from 2005 to 2009 was:

H.-G. Menzel (Chairman)	K.F. Eckerman	A.S. Pradhan
M. Balonov	J.D. Harrison	Y.-Z. Zhou
V. Berkovski	N. Ishigure	
W.E. Bolch	P. Jacob	
A. Bouville	J.L. Lipsztein	
G. Dietze	F. Paquet	

The membership of Committee 1 from 2009 to 2013 was:

J. Preston (Chairman)	C.R. Muirhead	D. Stram
T. Azizova	N. Nakamura	M. Tirmarche
R. Chakraborty	W. Rühm	R. Wakeford
S. Darby	S. Salomaa	P.-K. Zhou
J. Hendry	A.J. Sigurdson	
W.F. Morgan	F.A. Stewart	

The membership of Committee 2 from 2009 to 2013 was:

H.-G. Menzel (Chairman)	G. Dietze	J.L. Lipsztein
M. Balonov	K.F. Eckerman	J. Ma
D.T. Bartlett	A. Endo	F. Paquet
V. Berkovski	J.D. Harrison	N. Petoussi-Henss
W.E. Bolch	N. Ishigure	A.S. Pradhan
R. Cox	R. Leggett	

EXECUTIVE SUMMARY

(a) Epidemiological studies of occupational exposures of miners and domestic exposures of the public have provided strong and complementary evidence of the risks of lung cancer following inhalation of radon and its progeny. In the large cohorts of underground miners, annual occupational exposures were considered for the whole working period of each individual. Consequently, these studies are able to analyse dose–response relationships taking account of time-dependent modifying factors, such as age at exposure and time since exposure. The risk of lung cancer associated with domestic exposures to radon has been evaluated in a large number of case–control studies, requiring estimates of radon exposure in houses over a period of 30 years preceding lung cancer diagnosis. A weakness of such studies is that measurements made during the study period are assumed to apply throughout the whole period of exposure. An important strength, however, is that the residential studies often include detailed interviews so adjustments can be made, in the statistical analysis, for tobacco smoking as well as exposure to other potential lung carcinogens in the home or at work.

(b) In 1999, the BEIR VI report presented a comprehensive analysis of available miner cohorts from China, Czech Republic, USA, Canada, Sweden, Australia, and France (NRC, 1999). Recent studies of lung cancer in miners include relatively low concentrations of radon and its progeny, long duration of follow-up, and good-quality data for exposure of each individual (Tomášek et al., 2008; UNSCEAR, 2009). These results, consistent with previous analyses of combined miner studies, demonstrate significant associations between cumulative radon exposure and lung cancer mortality at levels of exposure as low as 50 working level months (WLM; i.e. 180 mJh/m^3). Based on lifetime excess absolute risk (LEAR) calculations, reference background rates from *Publication 103* (ICRP, 2007), and risk models derived from pooled analyses (NRC, 1999; Tomášek et al., 2008), a detriment-adjusted nominal risk coefficient of 5×10^{-4} per WLM [14×10^{-5} per (mJh/m^3)] is now recommended for radiological protection purposes. This nominal risk coefficient replaces the *Publication 65* value of 2.8×10^{-4} per WLM [8.0×10^{-5} per (mJh/m^3)].

(c) Three comprehensive publications have provided joint analyses of data from domestic case–control studies for Europe (Darby et al., 2005), North America (Krewski et al., 2005, 2006), and China (Lubin et al., 2004). Each joint analysis demonstrated an increased risk of lung cancer with increasing domestic radon concentration, considering exposures over a period of 30 years preceding diagnosis. The estimates of an increase of lung cancer per unit of concentration in the three joint analyses are very close and statistically compatible: the values obtained were 1.08, 1.10, and 1.13 per 100 Bq/m^3 from Europe, North America, and China, respectively. A combined estimate calculated for the studies in these three geographical areas was 1.09 per 100 Bq/m^3 (UNSCEAR, 2009). All of these results were obtained after adjustment for smoking habits. The slope of the linear exposure–response relationship increased slightly to 1.11 per 100 Bq/m^3 when analyses were restricted to cases and controls with more complete estimation of cumulated individual exposure (UNSCEAR, 2009).

(d) The joint analyses also adjusted for uncertainties associated with variations in radon concentration. For example, in the European pooled analysis (Darby et al., 2005), adjustment for measurement uncertainties markedly increased the estimate of relative risk from 1.08 to 1.16 per 100 Bq/m^3. Limiting the European analysis to those cases and controls with a relatively low annual exposure, there was evidence of an increased risk below 200 Bq/m^3. Analyses of the North American and Chinese studies were more variable and less statistically precise. It is concluded, however, that the residential studies provide consistent estimates of the risk of lung cancer and a basis for risk management related to low protracted radon exposures in homes, considering cumulative exposure over a period of at least 25 years.

(e) Although comparisons are complex, the cumulated excess absolute risk of lung cancer attributable to radon and its progeny estimated for residential exposures appears to be consistent with that obtained from miners at low levels of exposure.

(f) In the European pooled analysis of domestic exposures, a significant trend in the risk of lung cancer was observed among smokers, and also separately among non-smokers (Darby et al., 2006). Therefore, residential studies have demonstrated radon to be a lung carcinogen even in the absence of smoking, as shown previously in miner studies (Lubin et al., 1995). However, due to the dominant effect of tobacco use on lifetime risk of lung cancer, the excess absolute risk of lung cancer attributable to a given level of radon concentration is much higher among lifelong cigarette smokers than among non-smokers.

(g) The control of domestic exposures can be based directly on lung cancer risk estimates per unit exposure derived from epidemiological data; that is, in terms of radon concentrations in homes.

(h) However, for the purpose of control of occupational exposures using dose limits and constraints, estimates of dose per unit exposure are required. In *Publications 65 and 66* (ICRP, 1993, 1994), the effective dose per unit exposure to radon and its progeny was obtained using the so-called 'dose conversion convention'. This approach compared the detriment per unit exposure to radon and its progeny with the total detriment associated with unit effective dose, estimated largely on the basis of studies of Japanese atomic bomb survivors (ICRP, 1993). The values given were 5 mSv per WLM [1.4 mSv per (mJh/m^3)] for workers and 4 mSv per WLM [1.1 mSv per (mJh/m^3)] for members of the public.

(i) Doses from radon and its progeny can also be calculated using different dosimetric models. A review of published data on the effective dose per unit exposure to radon progeny obtained using dosimetric models is included as Annex B of this report. Values of effective dose range from about 6 to 20 mSv per WLM [1.7–5.7 mSv per (mJh/m^3)], with results using the Human Respiratory Tract Model (HRTM; ICRP, 1994) in the range from approximately 10 to 20 mSv per WLM [3–6 mSv per (mJh/m^3)] depending on the exposure scenario.

(j) ICRP has concluded that radon and its progeny should be treated in the same way as other radionuclides within the system of protection. That is, doses from radon and its progeny should be calculated using ICRP biokinetic and dosimetric models, including the HRTM and ICRP systemic models. In the near future, ICRP will provide dose coefficients per unit exposure to radon and its progeny for different reference conditions of domestic and occupational exposure, with specified equilibrium factors and aerosol characteristics. It should be recognised, however, that these dose coefficients will be larger by about a factor of two or more.

References

Darby, S., Hill, D., Auvinen, A., et al., 2005. Radon in homes and risk of lung cancer: collaborative analysis of individual data from 13 European case–control studies. Br. Med. J. 330, 223–227.

Darby, S., Hill, D., Deo, H., et al., 2006. Residential radon and lung cancer – detailed results of a collaborative analysis of individual data on 7148 persons with lung cancer and 14,208 persons without lung cancer from 13 epidemiological studies in Europe. Scand. J. Work Environ. Health 32 (Suppl. 1), 1–84.

ICRP, 1993. Protection against radon-222 at home and at work. ICRP Publication 65. Ann. ICRP 23(2).

ICRP, 1994. Human respiratory tract model for radiological protection. ICRP Publication 66. Ann. ICRP 24(1–3).

ICRP, 2007. The 2007 Recommendations of the International Commission on Radiological Protection. ICRP Publication 103. Ann. ICRP 37(2–4).

Krewski, D., Lubin, J.H., Zielinski, J.M., et al., 2005. Residential radon and risk of lung cancer. A combined analysis of 7 North American case–control studies. Epidemiology 16, 137–145.

Krewski, D., Lubin, J.H., Zielinski, J.M., et al., 2006. A combined analysis of North American case–control studies of residential radon and lung cancer. J. Toxicol. Environ. Health Part A 69, 533–597.

Lubin, J.H., Boice Jr., J.D., Edling, C., et al., 1995. Radon-exposed underground miners and inverse dose-rate (protraction enhancement) effects. Health Phys. 69, 494–500.

Lubin, J.H., Wang, Z.Y., Boice Jr., J.D., et al., 2004. Risk of lung cancer and residential radon in China: pooled results of two studies. Int. J. Cancer 109, 132–137.

NRC, 1999. Health Effects of Exposure to Radon. BEIR VI Report. National Academy Press, Washington, DC.

Tomášek, L., Rogel, A., Tirmarche, M., et al., 2008. Lung cancer in French and Czech uranium miners – risk at low exposure rates and modifying effects of time since exposure and age at exposure. Radiat. Res. 169, 125–137.

UNSCEAR, 2009. UNSCEAR 2006 Report. Annex E. Sources-to-Effects Assessment for Radon in Homes and Workplaces. United Nations, New York.

GLOSSARY

Case–control study
Type of epidemiological study in which a group of subjects with the disease of interest (e.g. cases with lung cancer) is compared with a group of subjects who are free of this disease (controls) but have similar characteristics (sex, attained age, etc.). This type of epidemiological design was most often used in indoor radon studies. For each individual, past exposures are estimated from measurements of radon concentration in current and previously occupied dwellings.

A nested case–control study is a specific type of case–control study, in which both cases and controls are extracted from a cohort study, aiming to obtain a more detailed evaluation than possible within the entire cohort.

Cohort study
Type of epidemiological study in which a population exposed to different levels of radon and its progeny is followed over time for the occurrence of diseases (including lung cancer). This type of epidemiological design was most often used in underground miner studies. The exposure in time was considered for each individual on an annual basis.

Detriment
Detriment is an ICRP concept. It reflects the total harm to health experienced by an exposed group and its descendants as a result of the group's exposure to a radiation source. Detriment is a multidimensional concept. Its principal components are the stochastic quantities: probability of attributable fatal cancer, weighted probability of attributable non-fatal cancer, weighted probability of severe heritable effects, and length of life lost if the harm occurs.

Dose conversion convention
This method, defined in *Publication 65* (ICRP, 1993), is used to relate exposure to radon progeny (expressed in WLM or Jh/m^3) to effective dose (expressed in mSv) on the basis of equal detriment.

Equilibrium equivalent concentration
The activity concentration of radon gas in equilibrium with its short-lived progeny that would have the same potential alpha energy concentration as the existing non-equilibrium mixture.

Equilibrium factor
The ratio of the equilibrium equivalent concentration to the radon gas concentration. In other words, the ratio of potential alpha energy concentration for the actual mixture of radon decay product to that which would apply at radioactive equilibrium.

Existing exposure situations
Exposure situations that already exist when a decision on control has to be taken. Such situations include exposure to natural background radiation, to naturally

occurring radioactive material, to residues in the environment resulting from operations that were not conducted within the Commission's system of protection, and to contaminated areas resulting from a nuclear accident or radiological event.

Human Respiratory Tract Model

Model used in *Publication 66* (ICRP, 1994) to evaluate the deposition and clearance of inhaled particles in the respiratory airways, as well as the resulting dose to the lung tissues.

Planned exposure situations

Planned exposure situations are situations involving the deliberate introduction and operation of sources. Planned exposure situations may give rise to exposures that are anticipated to occur (normal exposures) and to exposures that are not anticipated to occur (potential exposures).

Potential alpha energy concentration

The concentration of short-lived radon or thoron progeny in air in terms of the alpha energy emitted during complete decay from radon-222 progeny to lead-210, or from radon-220 progeny to lead-208, of any mixture of short-lived radon-222 or radon-220 in a unit volume of air.

Radon progeny

The decay products of radon-222, used in this report in the more limited sense of the short-lived decay products from polonium-218 to polonium-214. Radon progeny are sometimes referred to as 'radon decay products'.

Reference level

existing controllable exposure situations, this represents the level of dose or risk above which it is judged to be inappropriate to plan to allow exposures to occur, and below which optimisation of protection should be implemented. The chosen value for a reference level will depend upon prevailing circumstances of the exposure under consideration.

Risk

Risk relates to the probability or chance that an outcome (e.g. lung cancer) will occur. Terms relating to risk are listed below:

- Excess absolute risk

  An expression of risk based on the assumption that the excess risk from radiation exposure adds to the underlying (baseline) risk by an increment dependent on dose but independent of the underlying natural or background risk. In this report, lifetime excess absolute risk of lung cancer is computed.
- Relative risk

  The ratio of the incidence rate or the mortality rate from the disease of interest (e.g. lung cancer) in an exposed population to that in an unexposed population.
- Excess relative risk

  The rate of disease in an exposed population divided by the rate of disease in an unexposed population, minus 1. When studying a dose–response relationship, this is expressed as the excess relative risk per Gy or per Sv: (Relative risk − 1)/unit of exposure.

- Risk coefficient
  Increase of risk per unit exposure or per unit dose. In general, expressed as excess relative risk per WLM, per Jh/m^3, per 100 Bq/m^3, or per Sv.
- Risk model
  A model describing the variation of the risk coefficient as a function of modifying factors, such as time since exposure, attained age, or age at exposure. It may be related by a factor to the age-specific baseline risk (multiplicative) or added to the baseline risk (additive).
- Lifetime risk
  Risk cumulated by an individual up to a given age. The estimate used in the present report is the lifetime excess absolute risk associated with a chronic exposure scenario, expressed in number of deaths per 10 000 person-years per WLM (also sometimes denominated as the radiation excess induced death). In the present report, unless otherwise stated, the lifetime duration is 90 years as generally considered in ICRP publications, and the scenario is a constant low-level exposure to 2 WLM per year from 18 to 64 years of age, as proposed in *Publication 65* (ICRP, 1993).
- Detriment-adjusted risk
  The probability of the occurrence of a stochastic effect, modified to allow for the different components of the detriment in order to express the severity of the consequence(s).

Thoron progeny
  The decay products of radon-220, used herein in the more limited sense of the short-lived decay products from polonium-216 to polonium-212 or thallium-208.

Unattached fraction
  The fraction of the potential alpha energy concentration of short-lived radon progeny that is not attached to the ambient aerosol.

Upper reference levels
  Maximum values of exposure under which ICRP recommends national authorities to establish their own national reference levels.

Working level (WL)
  Any combination of the short-lived progeny of radon in one litre of air that will result in the emission of 1.3×10^5 MeV of potential alpha energy.
  1 WL = 2.08×10^{-5} J/m^3.

Working level month (WLM)
  The cumulative exposure from breathing an atmosphere at a concentration of 1 WL for a working month of 170 h.

Units

- Joules (J): $1\ J = 6.242 \times 10^{12}$ MeV
- Potential alpha energy concentration:
  for radon progeny:
  1 Bq/m^3 of radon at equilibrium = 3.47×10^4 MeV/m^3 = 5.56×10^{-9} J/m^3
  for thoron progeny:
  1 Bq/m^3 of thoron at equilibrium = 4.72×10^5 MeV/m^3 = 7.56×10^{-8} J/m^3
- Working level:
  1 WL = 1.3×10^8 MeV/m^3
  1 WL = 2.08×10^{-5} J/m^3
- Working level month:
  1 WLM = 3.54×10^{-3} Jh/m^3
  1 WLM = 6.37×10^5 Bqh/m^3 equilibrium equivalent concentration of radon
  1 WLM = 6.37×10^5/FBqh/m^3 of radon[a]
  1 Bq/m^3 of radon over 1 year = 4.4×10^{-3} WLM at home[b]
  1 Bq/m^3 of radon over 1 year = 1.26×10^{-3} WLM at work[b]
  1 WLM = 4.68×10^4 Bqh/m^3 equilibrium equivalent concentration of thoron

(a) F = equilibrium factor.
(b) Assuming 7000 h/year indoors or 2000 h/year at work, and F = 0.4 (ICRP, 1993).

References

ICRP, 1993. Protection against radon-222 at home and at work. ICRP Publication 65. Ann. ICRP 23(2).

ICRP, 1994. Human respiratory tract model for radiological protection. ICRP Publication 66. Ann. ICRP 24(1–3).

1. INTRODUCTION

(1) Radon-222 is a naturally occurring radioactive gas with a half-life of 3.8 days. It is formed as the decay product of radium-226 (half-life 1600 years), which is a member of the uranium-238 decay chain. Uranium and radium occur naturally in soil and rocks, and provide a continuous source of radon. Radon gas emanates from the earth's crust and, as a consequence, is present in the air outdoors and in all buildings, including workplaces. There is large variation in indoor air concentrations of radon, mainly due to the geology of the area and factors that affect the pressure differential between the inside and outside of the building, such as ventilation rates, heating within the building, and meteorological conditions.

(2) As radon is inert, nearly all of the gas that is inhaled is subsequently exhaled. However, radon-222 decays into a series of solid short-lived radioisotopes which, when inhaled, deposit within the respiratory tract. Due to their relatively short half-lives (<30 min), the radon progeny mainly decay in the lung before clearance can take place. Two of these short-lived progeny, polonium-218 and polonium-214, emit alpha particles, and it is the energy from these alpha particles that dominates dose to the lung and the associated risk of lung cancer.

(3) Radon has long been recognised as a cause of lung cancer, and was identified as a human lung carcinogen in 1986 by the World Health Organization (WHO, 1986; IARC, 1988). The main source of information on risks of radon-induced lung cancer has been epidemiological studies of underground miners (ICRP, 1993), and more recent studies have provided informative data on risks at lower levels of exposure (e.g. Lubin et al., 1997; NRC, 1999; EPA, 1999, 2003; Tomášek et al., 2008). In addition, recent combined analyses of data from case–control studies of lung cancer and residential radon exposures have demonstrated increased risk (Lubin et al., 2004; Darby et al., 2005, 2006; Krewski et al., 2006).

(4) The historical unit of exposure to radon progeny applied to the uranium mining environment is the working level month (WLM), which is related to the potential alpha energy concentration of its short-lived progeny. One WLM is defined as the cumulative exposure from breathing an atmosphere at a concentration of 1 working level (WL) for a working month of 170 h. A concentration of 1 WL is any combination of the short-lived radon progeny in one litre of air that will result in the emission of 1.3×10^5 MeV of alpha energy. One WLM is equivalent to 3.54×10^{-3} Jh/m^3 in SI units. Exposures can also be quantified in terms of the activity concentration of the radon gas in Bqh/m^3. The two units are related via the equilibrium factor (F), which is a measure of the degree of disequilibrium between radon and its short-lived progeny (1 WLM = 6.37×10^5 per FBqh/m^3; 1 Jh/m^3 = 1.8×10^8 per FBqh/m^3). Thus, an annual domestic exposure of 227 Bq/m^3 gives rise to 1 WLM, assuming occupancy of 7000 h/year and F = 0.4.

(5) A complication in the specification and control of doses and risks from radon has been that doses can be calculated in two ways: the so-called 'epidemiological' approach and the 'dosimetric' approach. *Publication 65* (ICRP, 1993) recommended an epidemiological approach in which the risk of fatal lung cancer per unit exposure to radon (in Jh/m^3 or WLM) was compared with the total risk, expressed as detriment,

per unit effective dose (in Sv). Hence, values of mSv (effective dose) per mJh/m^3 or WLM were obtained and referred to as the 'dose conversion convention'. Alternatively, various dosimetric models of the human respiratory tract, including the ICRP (1994) model, can be used to estimate equivalent dose to the lungs and effective dose per unit exposure to radon and its progeny. Given the uncertainties inherent in the estimation of risks from radiation exposure, and in the calculation of doses using dosimetric models, it is not surprising that the two approaches to calculating effective dose per unit exposure to radon have resulted in different values. In fact, the differences are remarkably small. However, the use of different values by different organisations, notably by ICRP (1993) and UNSCEAR (2000), suggests the need for clarification and the formulation of a consistent approach. ICRP now intends to treat radon and its progeny in the same way as other radionuclides, and to publish dose coefficients calculated using models for use within the ICRP system of protection.

(6) The present report considers epidemiological data on radon risks published since *Publication 65* (ICRP, 1993), focusing on studies involving low levels of protracted exposure. Results of pooled residential case–control studies are discussed in Chapter 2, and results of recent miner epidemiological studies with low exposures are discussed in Chapter 3. The miner data are used to recommend a revised estimate of lung cancer lifetime risk per unit exposure to radon progeny at low protracted levels of exposure to radon and its progeny. Annexes provide additional information on epidemiological results obtained from miner studies (Annex A), and review published results of dose per unit exposure to radon progeny and thoron progeny, calculated using dosimetric models of the human respiratory tract (Annex B).

1.1. References

Darby, S., Hill, D., Auvinen, A., et al., 2005. Radon in homes and risk of lung cancer: collaborative analysis of individual data from 13 European case–control studies. Br. Med. J. 330, 223–227.

Darby, S., Hill, D., Deo, H., et al., 2006. Residential radon and lung cancer – detailed results of a collaborative analysis of individual data on 7148 persons with lung cancer and 14,208 persons without lung cancer from 13 epidemiological studies in Europe. Scand. J. Work Environ. Health 32 (Suppl. 1), 1–84.

EPA, 1999. Proposed Methodology for Assessing Risks from Indoor Radon Based on BEIR VI. Office of Radiation and Indoor Air, United States Environmental Protection Agency, Washington, DC.

EPA, 2003. Assessment of Risks from Radon in Homes. Publication EPA 402-R-03-003. Office of Air and Radiation, United States Environmental Protection Agency, Washington, DC.

IARC, 1988. Monographs on the Evaluation of Carcinogenic Risk to Humans: Man-made Fibres and Radon. IARC 43. International Agency for Research on Cancer, Lyon.

ICRP, 1993. Protection against radon-222 at home and at work. ICRP Publication 65. Ann. ICRP 23(2).

ICRP, 1994. Human respiratory tract model for radiological protection. ICRP Publication 66. Ann. ICRP 24(1–3).

Krewski, D., Lubin, J.H., Zielinski, J.M., et al., 2006. A combined analysis of North American case–control studies of residential radon and lung cancer. J. Toxicol. Environ. Health Part A 69, 533–597.

Lubin, J.H., Tomášek, L., Edling, C., et al., 1997. Estimating lung cancer mortality from residential radon using data for low exposures of miners. Radiat. Res. 147, 126–134.

Lubin, J.H., Wang, Z.Y., Boice Jr., J.D., et al., 2004. Risk of lung cancer and residential radon in China: pooled results of two studies. Int. J. Cancer 109, 132–137.

NRC, 1999. Health Effects of Exposure to Radon. BEIR VI Report. National Academy Press, Washington, DC.

Tomášek, L., Rogel, A., Tirmarche, M., et al., 2008. Lung cancer in French and Czech uranium miners – risk at low exposure rates and modifying effects of time since exposure and age at exposure. Radiat. Res. 169, 125–137.

UNSCEAR, 2000. Sources and Effects of Ionizing Radiation. UNSCEAR 2000 Report to the General Assembly with Scientific Annexes. United Nations, New York.

WHO, 1986. Indoor Air Quality Research: Report on a WHO Meeting, 27–31 August 1984, Stockholm. World Health Organization, Copenhagen.

2. EPIDEMIOLOGY OF THE RISK OF LUNG CANCER ASSOCIATED WITH RESIDENTIAL EXPOSURES TO RADON AND ITS PROGENY

2.1. Introduction

(7) In 1988, the International Agency for Research on Cancer classified radon as a human lung carcinogen, based on a review of evidence from experimental data on animals and from epidemiological studies of underground miners exposed to relatively high concentrations of radon and its progeny. This report is focused on those epidemiological studies able to provide information on the dose–response relationship with the risk of lung cancer at relatively low annual exposures to radon and its progeny. Studies that include both individual exposure assessment and individual assessment of potential confounding factors or cofactors, such as tobacco use, are given particular emphasis. Ecological studies of cancer rates and average exposure per country or per region do not provide individual exposure data and are not considered; these studies are unable to provide reliable information on risk, and are limited due to the unknown effect of confounding factors, including smoking, and the unknown influence of population movement into and out of the study areas (WHO, 1996; NRC, 1999).

(8) The applicability of studies of underground miners to estimate radon-induced lung cancer for residential concentrations of radon has been an important source of uncertainty over the last 20 years. A variety of factors need to be considered in this extrapolation from mines to homes. These include: the linearity of the dose–response relationship; any differences between risks for adult males and the general population, which includes women and children; the difference in other environmental exposures which may include arsenic, quartz, and diesel exhaust amongst others; different F values for radon and its short-lived progeny; and different breathing rates.

(9) Due to the desirability of having direct information on risks associated with domestic radon concentrations, a large number of residential epidemiological studies were launched in the late 1980s and early 1990s. There was also an awareness that pooling of data may be required to provide the statistical power to demonstrate a significant risk at residential exposure concentrations (Lubin and Boice, 1997). Reliable estimates of individual exposure conditions over long periods of time were an important prerequisite of the epidemiological studies, with long-term radon measurement in the current and previous homes of each individual. Individuals' habits and ventilation conditions in dwellings also had to be considered.

2.2. Studies published since 1990

(10) This section considers analytical epidemiological studies that have included at least 200 cases of lung cancer, as well as long-term domestic radon measurements. Table 2.1 summarises 20 case–control studies published between 1990 and 2006. More extensive details are available elsewhere (UNSCEAR, 2009).

Table 2.1. Residential radon case–control studies and one cohort study with more than 200 cases of lung cancer published between 1990 and 2006.

Reference	Region	Population	No. of cases/controls	Measurement period	Relative risk per 100 Bq/m^3	95% CI
Schoenberg et al. (1990)	USA (New Jersey)	Females	480 cases, 442 controls	1 year	1.49	0.89–1.89
Blot et al. (1990)	China (Shenyang)	Females	308 cases, 356 controls	1 year	0.95	Undefined–1.08
Pershagen et al. (1992)	Sweden	Females	201 cases, 378 controls	1 year	1.16	0.89–1.92
Pershagen et al. (1994)	Sweden	Males and females	1281 cases, 2576 controls	3 months	1.10	1.01–1.22
Letourneau et al. (1994)	Canada	Males and females	738 cases, 738 controls	1 year	0.98	0.87–1.27
Alavanja et al. (1994)	USA (Missouri)	Females, never-smokers	538 cases, 1183 controls	1 year	1.08	0.95–1.24
Auvinen et al. (1996)	Finland	Males and females	517 cases, 517 controls	1 year	1.11	0.94–1.31
Ruosteenoja et al. (1996)	South Finland	Males	318 cases, 1500 controls	2 months	1.80	0.90–3.50
Darby et al. (1998)	UK	Males and females	982 cases, 3185 controls	6 months	1.08	0.97–1.20
Alavanja et al. (1999)	USA (Missouri)	Females	477 cases, 516 controls	1 year	1.27	0.88–1.53
			387 cases, 473 controls		1.3	1.07–2.93
Field et al. (2000)	USA (Iowa)	Females	413 cases, 614 controls	1 year	1.24	0.95–1.92
Kreienbrock et al. (2001)	Germany (West)	Males and females	1449 cases, 2297 controls	1 year	0.97	0.82–1.14
Lagarde et al. (2001)	Sweden	Never-smokers	436 cases, 1649 controls	3 months	1.10	0.96–1.38
Wang et al. (2002)	China (Gansu)	Males and females	768 cases, 1659 controls	1 year	1.19	1.05–1.47
Kreuzer et al. (2003)	Germany (East)	Males and females	1192 case, 1640 controls	1 year	1.08	0.97–1.20
Baysson et al. (2004)	France	Males and females	486 cases, 984 controls	6 months	1.04	0.99–1.11
Bochicchio et al. (2005)	Italy	Males and females	384 cases, 404 controls	6 + 6 months	1.14	0.89–1.46
Sandler et al. (2006)	USA (Connecticut + Utah-South Idaho)	Males and females	1474 cases, 1811 controls	1 year	1.01	0.79–1.21
Tomášek et al. (2001)	Czech Republic	Males and females	173 cases in a cohort of 12,000 inhabitants	1 year	1.10	1.04–1.17

CI, confidence interval.

28

(11) In most of the studies, year-long measurements of radon and its decay products were made using standard methodologies in order to integrate any variations in the specific conditions of the dwellings, and any climatic and seasonal changes. Most measurements were of concentrations in air using alpha track detectors. In a few studies, glass-based retrospective detectors were also used.

(12) A number of European studies were designed with the intention of conducting a pooled analysis (see Section 2.3). Considerable efforts were made to have comparable protocols before starting studies in different countries. They were all case–control studies, with face-to-face interviews performed whenever possible for both the cases (patients with lung cancer) and the controls (hospital controls or controls from the general population). The same detailed questionnaire was used to analyse the risk of lung cancer in relation to domestic radon exposure, adjusting for tobacco consumption, occupational exposures, and indicators of socio-economic status. These studies provide information on the risk of lung cancer from radon for smokers and non-smokers, and allow adjustment to be made relating to years as a smoker, age at onset of smoking, years since smoking cessation, and average number of cigarettes smoked per day. Several large case–control studies were also conducted in Canada and the USA, as well as two studies in China (one in Shenyang and one in Gansu).

(13) The studies listed in Table 2.1 evaluated the association between lung cancer and domestic radon exposure. Results are presented in terms of the relative risk per 100 Bq/m^3 averaged for most studies over 20–30 years prior to diagnosis of lung cancer. Two studies only considered never-smokers; most studies considered males and females, and smokers and non-smokers. Risks of radon exposure are adjusted for smoking habits, and in several studies are also adjusted for occupational exposures known to be potential lung carcinogens (e.g. asbestos). Most of the studies (17 out of 20 independent studies) reported a positive trend in the risk of lung cancer with increasing exposure, but few of the trends were significant. A few studies did not find a positive trend. Considered alone, each study had low statistical power and provided an estimate of the risk per unit of exposure with a large confidence interval (CI). Most studies only included a small number of cases of lung cancer that were never-smokers; as such, they were limited in evaluating associations between radon decay products and lung cancer in non-smoking populations.

(14) In most studies, the radon concentration could not be measured in some residences (e.g. if the house had been demolished). In these cases, radon concentrations had to be estimated for the purposes of the statistical analyses. Even when the radon concentration had been measured in a home, the measurements were subject to uncertainty in the sense that repeated measurements in the same residence and in the same period showed a high variability of radon levels. The inability to detect an association in many individual studies may have been due to poor retrospective radon exposure assessment, and/or to there being very few cases and controls living in residences with radon concentrations over 200 Bq/m^3. In several studies, the average time-weighted radon concentrations in homes occupied by cases and controls were low, and only a few studies [e.g. in the Czech Republic, Finland, France,

Sweden, and Gansu (China)] included persons living in houses with levels of exposure above 400 Bq/m^3.

2.3. Pooled studies

(15) Since 2000, several joint analyses have been published, integrating the basic individual data from cases and controls, and applying a standard methodology in defining selection criteria and statistical analysis. It is noted that several informative meta-analyses of radon studies have been conducted but did not have the strengths of these pooled analyses which handle individual data in the same manner (Lubin and Boice, 1997; NRC, 1999; UNSCEAR, 2009). Three joint analyses have been conducted based on data from Europe (Darby et al., 2005), North America (Krewski et al., 2005, 2006), and China (Lubin et al., 2004) (Table 2.2). Each joint analysis showed evidence of the risk of lung cancer increasing with cumulated domestic exposure to radon. The exposure period considered was at least 30 years prior to diagnosis of lung cancer for the North American and Chinese joint analysis, and 35 years prior to diagnosis for the European joint analysis. In each analysis, the radon concentrations estimated for the 5 years prior to diagnosis were not considered since a minimum lag time of 5 years was assumed from lung cancer induction to diagnosis based on data from studies of underground miners (NRC, 1999). In consequence, the estimated risk per unit of exposure is based on a time-weighted average exposure for a window period 5–30 years prior to diagnosis (5–34 years for the European pooled analysis). The estimates of the increased risk of lung cancer per unit exposure in the three joint analyses are very close and statistically compatible (Table 2.2): the values obtained were 1.08, 1.10, and 1.13 per 100 Bq/m^3 for Europe, America, and China, respectively. The combined estimate for Europe, North America, and China was 1.09 per 100 Bq/m^3 (UNSCEAR, 2009).

(16) The relative risk of lung cancer was shown to be increased among both smokers and non-smokers. In the European joint analysis, the estimated relative risk per 100 Bq/m^3 was 1.11 (95% CI 1.00–1.28) for lifelong non-smokers; in the joint North American study, the relative risk for non-smokers was of the same level (1.10) but was not significant (95% CI 0.91–1.42).

(17) It is noteworthy that the slope of the linear exposure–response relationship increased when analyses were restricted to individuals with the most precise estimates of cumulative exposure (e.g. when restricted to individuals resident in the same house

Table 2.2. Pooled analyses of case–control studies of residential exposure to radon and lung cancer, based on measured radon concentrations.

Joint analysis	No. of studies included	No. of cases	No. of controls	Relative risk per 100 Bq/m^3 (95% CI)
European (Darby et al., 2006)	13	7148	14,208	1.08 (1.03–1.16)
North American (Krewski et al., 2006)	7	3662	4966	1.10 (0.99–1.26)
Chinese (Lubin et al., 2004)	2	1050	1995	1.13 (1.01–1.36)

CI, confidence interval.

for the previous 20 years). In the North American study (Krewski et al., 2005, 2006), analysis restricted in terms of residential stability (i.e. only one or two houses occupied in the 5–30 years preceding diagnosis) and completeness of radon monitoring (measurements for at least 20 years of the considered period) resulted in an estimated relative risk of 1.18 per 100 Bq/m^3. In the Chinese analysis (Lubin et al., 2004), when considering only those subjects who had lived in their current homes for 30 years or more, the estimated relative risk was 1.32 (95% CI 1.07–1.91). According to the UNSCEAR 2006 report, for all three joint analyses combined, the slope of the linear exposure–response relationship was 1.11 per 100 Bq/m^3 when the analyses focused on those cases and controls with more precise estimates of cumulated individual exposure (UNSCEAR, 2009).

(18) The joint analyses also tried to take account of uncertainties associated with variations in exposure (Fearn et al., 2008). In the European pooled analysis (Darby et al., 2005, 2006), the estimated relative risk increased from 1.08 to 1.16 per 100 Bq/m^3 after taking account of random uncertainties in radon measurements.

(19) Limiting the European analysis to those cases and controls with a relatively low annual exposure, there is convincing evidence of an increased risk of lung cancer for those exposed to levels below 200 Bq/m^3 (Darby et al., 2006).

(20) One of the strengths of these joint analyses is that efforts were made to collect detailed past smoking habits on the basis of direct interviews in most studies, and each analysis included adjustment for smoking. For the European pooled analysis (Darby et al., 2005, 2006), a negative correlation was demonstrated between residential radon exposure and smoking, meaning that failure to take account of smoking would have biased the estimates of risks from radon towards the null. The relative risk of lung cancer per 100 Bq/m^3 was 1.02 when stratified by study, region, age, and sex, but not smoking. This estimate increased to 1.05 when also stratified for smoking using seven categories (never-smokers; current cigarette smokers of <15, 15–24 or $\geqslant$25 cigarettes per day; ex-smokers of <10 years' or $\geqslant$10 years' duration; and unknown). A further increase to 1.08 was observed when current smokers were further stratified by age at smoking onset, and ex-smokers were stratified by the number of cigarettes smoked.

(21) In conclusion, the joint analyses of the risk of lung cancer from residential radon exposures show an increase in risk of at least 8% per 100 Bq/m^3, considering a period of exposure from 5 to 30–35 years prior to diagnosis of lung cancer. When restricting the analysis to those with presumably more precise exposure measurements, the observed risk is increased in each of the joint analyses. The European pooled analysis reported an increase in excess relative risk (ERR) of 16% per 100 Bq/m^3 when uncertainties in the measured radon activity concentrations were considered. This value may be considered to be a reasonable estimate of the risk associated with relatively low and prolonged radon exposures in homes, considering exposure over a period of 25–30 years.

(22) When the analysis is limited to lifelong non-smokers, a significant positive trend is still observed in the European pooled analysis, based on a large number

of cases of lung cancer: 268 in men and 616 in women, and more than 5000 controls (Darby et al., 2006).

(23) On the basis of the results of the European pooled analysis, the cumulative risk of lung cancer up to 75 years of age is estimated for lifelong non-smokers as 0.4%, 0.5%, and 0.7% for radon activity concentrations of 0, 100, and 400 Bq/m^3, respectively. Lifelong cigarette smokers have a baseline risk of lung cancer that is about 25 times higher than that for non-smokers. The lifetime cumulative risks of lung cancer for lifelong smokers by 75 years of age are close to 10%, 12%, and 16% for radon activity concentrations of 0 (theoretical non-exposed situation), 100, and 400 Bq/m^3, respectively, and reflect the dominant effect of tobacco use on lifetime risk of lung cancer with or without the contribution of radon.

(24) A 'world pooled' analysis is in progress under the co-ordination of Sarah Darby (Oxford University), considering more than 13,700 cases of lung cancer from 25 studies. It will include three supplementary studies: one from Russia (Urals) and two from North America (Massachusetts and New Jersey). Results from this large joint analysis are expected in the near future. They may provide better adjustments for cofactors, but as the dominant studies included are considered here in the three separate joint analyses from Europe, North America, and China, the overall conclusion is expected to be the same: clear evidence of an increased relative risk of lung cancer related to radon exposure cumulated in houses over a residence period of at least 30 years prior to diagnosis.

2.4. References

Alavanja, M.C., Brownson, R.C., Lubin, J., et al., 1994. Residential radon exposure and lung cancer among nonsmoking women. J. Natl. Cancer Inst. 86, 1829–1837.

Alavanja, M.C., Lubin, J.H., Mahaffey, J.A., Brownson, R.C., 1999. Residential radon exposure and risk of lung cancer in Missouri. Am. J. Public Health 89, 1042–1048.

Auvinen, A., Mäkeläinen, I., Hakama, M., et al., 1996. Indoor radon exposure and risk of lung cancer: a nested case–control study in Finland. J. Natl. Cancer Inst. 88, 966–972.

Baysson, H., Tirmarche, M., Tymen, G., 2004. Indoor radon and lung cancer in France. Epidemiology 15, 709–716.

Blot, W.J., Xu, Z.Y., Boice Jr., J.D., et al., 1990. Indoor radon and lung cancer in China. J. Natl. Cancer Inst. 82, 1025–1030.

Bochicchio, F., Forastiere, F., Farchi, S., et al., 2005. Residential radon exposure, diet and lung cancer: a case–control study in a Mediterranean region. Int. J. Cancer 114, 983–991.

Darby, S., Whitley, E., Silcocks, P., et al., 1998. Risk of lung cancer associated with residential radon exposure in south-west England: a case–control study. Br. J. Cancer 78, 394–408.

Darby, S., Hill, D., Auvinen, A., et al., 2005. Radon in homes and risk of lung cancer: collaborative analysis of individual data from 13 European case–control studies. Br. Med. J. 330, 223–227.

Darby, S., Hill, D., Deo, H., et al., 2006. Residential radon and lung cancer – detailed results of a collaborative analysis of individual data on 7148 persons with lung cancer and 14,208 persons without lung cancer from 13 epidemiological studies in Europe. Scand. J. Work Environ. Health 32 (Suppl. 1), 1–84.

Fearn, T., Hill, D.C., Darby, S.C., 2008. Measurement error in the explanatory variable of a binary regression: regression calibration and integrated conditional likelihood in studies of residential radon and lung cancer. Stat. Med. 27, 2139–2176.

Field, R.W., Steck, D.J., Smith, B.J., et al., 2000. Residential radon gas exposure and lung cancer: the Iowa Radon Lung Cancer Study. Am. J. Epidemiol. 151, 1091–1102.

Kreienbrock, L., Kreuzer, M., Gerken, M., et al., 2001. Case–control study on lung cancer and residential radon in western Germany. Am. J. Epidemiol. 53, 42–52.

Kreuzer, M., Heinrich, J., Wölke, G., et al., 2003. Residential radon and risk of lung cancer in Eastern Germany. Epidemiology 14, 559–568.

Krewski, D., Lubin, J.H., Zielinski, J.M., et al., 2005. Residential radon and risk of lung cancer. A combined analysis of 7 North American case–control studies. Epidemiology 16, 137–145.

Krewski, D., Lubin, J.H., Zielinski, J.M., et al., 2006. A combined analysis of North American case–control studies of residential radon and lung cancer. J. Toxicol. Environ. Health Part A 69, 533–597.

Lagarde, F., Axelsson, G., Damber, L., et al., 2001. Residential radon and lung cancer among never-smokers in Sweden. Epidemiology 12, 396–404.

Letourneau, E.G., Krewski, D., Choi, N.W., et al., 1994. Case–control study of residential radon and lung cancer in Winnipeg, Manitoba, Canada. Am. J. Epidemiol. 140, 310–322.

Lubin, J.H., Boice Jr., J.D., 1997. Lung cancer risk from residential radon: meta-analysis of eight epidemiologic studies. J. Natl. Cancer Inst. 89, 49–57.

Lubin, J.H., Wang, Z.Y., Boice Jr., J.D., et al., 2004. Risk of lung cancer and residential radon in China: pooled results of two studies. Int. J. Cancer 109, 132–137.

NRC, 1999. Health Effects of Exposure to Radon. BEIR VI Report. National Academy Press, Washington, DC.

Pershagen, G., Liang, Z.H., Hrubec, Z., et al., 1992. Residential radon exposure and lung cancer in Swedish women. Health Phys. 63, 179–186.

Pershagen, G., Akerblom, G., Axelson, O., et al., 1994. Residential radon exposure and lung cancer in Sweden. N. Engl. J. Med. 330, 159–164.

Ruosteenoja, E., Mäkeläinen, I., Rytömaa, T., et al., 1996. Radon and lung cancer in Finland. Health Phys. 71, 185–189.

Sandler, D.P., Weinberg, C.R., Shore, D.L., et al., 2006. Indoor radon and lung cancer risk in Connecticut and Utah. J. Toxicol. Environ. Health A 69, 633–654.

Schoenberg, J.B., Klotz, J.B., Wilcox, H.B., et al., 1990. Case–control study of residential radon and lung cancer among New Jersey women. Cancer Res. 50, 6520–6524.

Tomášek, L., Kunz, E., Müller, T., et al., 2001. Radon exposure and lung cancer risk – Czech cohort study on residential radon. Sci. Total Environ. 272, 43–51.

UNSCEAR, 2009. UNSCEAR 2006 Report, Annex E. Sources-to-Effects Assessment for Radon in Homes and Workplaces. United Nations, New York.

Wang, Z., Lubin, J.H., Wang, L., et al., 2002. Residential radon and lung cancer risk in a high-exposure area of Gansu Province, China. Am. J. Epidemiol. 155, 554–564.

WHO, 1996. Radon. World Health Organization, Copenhagen.

3. EPIDEMIOLOGY OF THE RISK OF LUNG CANCER ASSOCIATED WITH EXPOSURE TO RADON AND ITS PROGENY IN UNDERGROUND MINES

3.1. Review of results since *Publication 65*

(25) *Publication 65* (ICRP, 1993) estimated the risk of lung cancer mortality from radon exposure on the basis of studies on seven cohorts of miners [Colorado (USA), Ontario (Canada), New Mexico (USA), Beaverlodge (Canada), Western Bohemia (Czech Republic), CEA-COGEMA (France) and Malmberget (Sweden)] (Table A.1 in Annex A). The total number of miners was 31,486. The weighted average of the ERR per 100 WLM for these studies was 1.34 (95% CI 0.82–2.13). This ERR coefficient applied to a follow-up period of 20 years, taking into account a lag time (minimum latency) of 5 years, i.e. radon results for exposures experienced 5 years prior to death from lung cancer (or comparable date for other miners) were excluded from the analyses. A model was derived, taking account of the modifying effects of age at exposure and time since exposure (TSE) (ICRP, 1993).

(26) A comprehensive analysis of epidemiological results based on 11 cohorts of radon-exposed miners was published in 1994 (Lubin et al., 1994). In comparison with the *Publication 65* report, results for some cohorts were updated [Colorado (USA), Ontario (Canada), Beaverlodge (Canada), Western Bohemia (Czech Republic) and Malmberget (Sweden)] and other cohorts were added [Yunnan (China), Newfoundland (Canada), Port Radium (Canada) and Radium Hill (Australia)]. This analysis gave an ERR per 100 WLM of 0.49 (95% CI 0.2–1.0) (Lubin et al., 1994). After some minor updates of the same 11 cohorts, a new joint analysis was published in the BEIR VI report (NRC, 1999). This joint analysis relied on a total of 60,606 miners, with a total of 2674 deaths from lung cancer (Table A.2 in Annex A). The estimated combined ERR per 100 WLM was 0.59, assuming an exposure lag time of 5 years. Two models were derived, taking account of modifying effects of attained age and TSE, as well as either duration of exposure or mean rate of exposure. Analyses on restricted ranges of cumulative exposure of less than 100 or 50 WLM were also performed (NRC, 1999).

(27) Since the BEIR VI report (NRC, 1999), new results have been published for the West Bohemian cohort (uranium mines) and the North Bohemian cohort (tin mines) in the Czech Republic (Tomášek and Placek, 1999; Tomášek, 2002; Tomášek et al., 2003; Tomášek and Zarska, 2004); the Newfoundland cohort (fluorspar mines) (Villeneuve et al., 2007) and the Eldorado cohort (including workers from Port Radium and Beaverlodge) (Howe, 2006; Lane et al., 2010) in Canada; the Colorado Plateau cohort in the USA (Schubauer-Berigan et al., 2009); the Wismut uranium mines in Germany (Kreuzer et al., 2002, 2008, 2010; Grosche et al., 2006; Schnelzer et al., 2010; Walsh et al., 2010a,b); and the CEA-COGEMA mines in France (Rogel et al., 2002; Laurier et al., 2004; Vacquier et al., 2008, 2009).

(28) The UNSCEAR 2006 report (UNSCEAR, 2009) provided a comprehensive review of available epidemiological results from nine studies [the New Mexico (USA) and Australian studies were not included], including a total of more than

126,000 miners (Table A.3 in Annex A). The weighted mean average ERR per 100 WLM was 0.59 (95% CI 0.35–1.0) (UNSCEAR, 2009).

(29) Since the UNSCEAR 2006 report (UNSCEAR, 2009), the results of a joint analysis of the Czech and French miner cohorts have been published. This analysis included 10,100 miners with a relatively long follow-up (mean of approximately 24 years) and relatively low levels of cumulative exposure (mean of 46.8 WLM). The estimated ERR per 100 WLM was 1.6 (95% CI 1.0–2.3) (Tirmarche et al., 2003; Tomášek et al., 2008).

(30) Although other miner studies have been published, they are generally not included here or in other comprehensive summaries as they provide little to no quantitative information on the relationship between radon and cancer risk.

3.2. Summary of estimates of excess relative risk per 100 working level months

(31) The results from combined analyses summarised in Table 3.1 are presented as simple linear estimates of ERR per 100 WLM. They apply across the whole population of the cohorts under consideration, but do not reflect variations in risk between or within the cohorts. Some characteristics of the cohorts may explain variations in the estimated ERR per 100 WLM, including duration of follow-up, attained age, duration of work, exposure levels, and background rates of lung cancer. It is important, therefore, to consider such factors in assessment of the risk associated with exposure to radon and its progeny. Nevertheless, the three large-scale analyses that summarise most currently available information (Lubin et al., 1994; NRC, 1999; UNSCEAR, 2009) provide estimates of the association between cumulative WLM exposure and risk of lung cancer that are highly concordant.

(32) All of the combined analyses and some of the individual studies demonstrate a modifying effect of TSE and, to a lesser extent, attained age (ICRP, 1993; Lubin et al., 1994; NRC, 1999; Howe, 2006; Tomášek et al., 2008). An inverse exposure–rate effect (or protraction enhancement effect) has also been observed in most analyses (Lubin et al., 1994; NRC, 1999), although such a modifying effect is not seen at low levels of cumulative WLM exposure (Lubin et al., 1995; Tomášek et al., 2008), or was no more evident using improved individual dosimetric data (Vacquier et al., 2009). Models have been developed to combine the modifying effects of TSE, age, and exposure rate. Two models were proposed in the BEIR VI report:

Table 3.1. Summary of excess relative risk (ERR) per 100 working level months (WLM) published from combined analyses of miner studies.

Reference	No. of cohorts	No. of miners	Person-years	ERR per 100 WLM	SE	95% CI
ICRP (1993)	7	31,486	635,022	1.34		0.82–2.13
Lubin et al. (1994)	11	60,570	908,903	0.49		0.20–1.00
NRC (1999)	11	60,705	892,547	0.59	1.32	
UNSCEAR (2009)	9	125,627	3,115,975	0.59		0.35–1.00
Tomášek et al. (2008)	2	10,100	248,782	1.60		1.00–2.30

SE, standard error; CI, confidence interval.

Table 3.2. Estimates of the excess relative risk (ERR) per working level month (WLM) based on subgroups with low levels of exposure and low exposure rate.

Reference	Model	Exposure	ERR per 100 WLM	95% CI
NRC (1999)	BEIR VI restricted range	<100 WLM	0.81	0.30–1.42
NRC (1999)	BEIR VI restricted range	<50 WLM	1.18	0.20–2.53
NRC (1999)	BEIR VI TSE–age–concentration model	Rate <0.5 WL	3.41[a]	–
Howe (2006)	Beaverlodge	Mean 85 WLM	0.96	0.56–1.56
Kusiak et al. (1993)	Ontario	Mean 31 WLM	0.89	0.5–1.5
Vacquier et al. (2008)	French cohort, employed after 1956	Mean 17 WLM	2.0	0.91–3.65
Tomášek et al. (2008)	Joint Czech–French cohort[b]	Mean 47 WLM	2.7[a]	1.7–4.3

TSE, time since exposure; CI, confidence interval; WL, working level.

[a] For an attained age of 55–64 years at 15–24 years following exposure.

[b] Restricted to miners with measured radon exposures.

the TSE–age–concentration model and the TSE–age–duration model (NRC, 1999). These models provide risk coefficients for different windows of cumulative exposure, with additional modifying effects of age and concentration/duration based on categorical variables. An alternative approach has been proposed in the joint analysis of the Czech and French cohorts, modelling the risk associated with cumulative radon exposure and integrating the modifying effects of TSE and attained age as continuous variables (Tomášek et al., 2008).

(33) For current radiation protection purposes, the most relevant results from miner studies are those derived for populations with low levels of cumulative exposure, long duration of follow-up, and good-quality data. In general, the ERR per 100 WLM estimated from cohorts with a low level of exposure (e.g. the Ontario, Beaverlodge, and French cohorts) are higher than those estimated from cohorts with high levels of cumulative exposure, although the CIs are broader (Table A.3 in Annex A). Some publications have provided estimates based on analyses on restricted ranges of exposure (Lubin et al., 1997). In the BEIR VI report, such analyses resulted in estimated ERRs per 100 WLM of 0.81 and 1.18 below 100 WLM and 50 WLM, respectively (NRC, 1999). In addition, coefficients corresponding to low exposure rates can be obtained from models that take account of modifying factors. In the BEIR VI report, an ERR per 100 WLM of 3.41 was obtained for low exposure rates below 0.5 WL (TSE–age–concentration model, for an attained age of 55–64 years and at 15–24 years following exposure) (NRC, 1999). Recent analyses from the French and Czech cohorts have provided risk estimates associated with low levels of exposure and reasonably good-quality exposure assessment ('measured exposures'), with values of ERR per 100 WLM varying between 2.0 and 3.4 (Tomášek et al., 2008; Vacquier et al., 2008). A summary of these risk estimates is presented in Table 3.2, demonstrating significant associations between cumulative radon exposure and lung cancer mortality at low levels of cumulative exposure.

3.3. Risk of lung cancer from radon and smoking

(34) Although smoking is by far the strongest risk factor for lung cancer, most studies of underground miners could not take account of smoking habits. Several studies have partial smoking data, including the Yunnan cohort (China), the Colorado Plateau cohort (USA), the Newfoundland fluorspar miner cohort (Canada), the Sweden cohort, the New Mexico cohort (USA), and the Radium Hill cohort (South Australia). Case–control studies among miners have also been conducted to investigate the interaction between radon exposure and smoking on risk of lung cancer (Qiao et al., 1989; Lubin et al., 1990; L'Abbé et al., 1991; Thomas et al., 1994; Yao et al., 1994; Brüske-Hohlfeld et al., 2006; Leuraud et al., 2007; Amabile et al., 2009). More information on the risk of lung cancer associated with both radon and cigarette smoking should be available in the future as new datasets from cohort and case–control studies are currently under development in Canada (Ontario cohort) and Europe (Czech, German, and French cohorts) (Tirmarche et al., 2010).

(35) Considering currently available data, the results indicate that the relationship between lung cancer mortality and radon exposure persists when account is taken of smoking habits. The analyses conducted for the BEIR VI report demonstrated a sub-multiplicative interaction between radon exposure and smoking status (NRC, 1999). In the Newfoundland fluorspar miner cohort, the ERR per 100 WLM was not significantly different between never-smokers (ERR per 100 WLM of 0.42) and ever-smokers (ERR per 100 WLM of 0.48). However, a significant increase in the ERR per 100 WLM with increasing number of cigarettes smoked daily was noted (Villeneuve et al., 2007). In a recent French nested case–control study, the ERR for lung cancer related to cumulative radon exposure, adjusted for smoking, was 0.85 per 100 WLM (Leuraud et al., 2007). Adjustment for smoking only led to marginal changes in the risk of lung cancer associated with radon (Leuraud et al., 2007; Schnelzer et al., 2010). Tirmarche et al. (2003) concluded that presently available models derived from cohort studies of underground miners that do not take account of smoking status appear to be acceptable for estimation of the risk of lung cancer associated with radon in a population including both smokers and non-smokers. When the smoking status is known, the estimated ERR is generally larger (even if not significantly) among non-smokers than smokers (Lubin et al., 1994; Tomášek, 2002).

3.4. References

Amabile, J.C., Leuraud, K., Vacquier, B., et al., 2009. Multifactorial study of the risk of lung cancer among French uranium miners: radon, smoking and silicosis. Health Phys. 97, 613–621.

Brüske-Hohlfeld, I., Rosario, A.S., Wölke, G., et al., 2006. Lung cancer risk among former uranium miners of the WISMUT Company in Germany. Health Phys. 90, 208–216.

Grosche, B., Kreuzer, M., Kreisheimer, M.A., 2006. Lung cancer risk among German male uranium miners: a cohort study, 1946–1998. Br. J. Cancer 95, 1280–1287.

Howe, G.R., 2006. Updated Analysis of the Eldorado Uranium Miner's Cohort: Part I of the Saskatchewan Uranium Miner's Cohort Study. RSP-0205. Columbia University, New York.

ICRP, 1993. Protection against radon-222 at home and at work. ICRP Publication 65. Ann. ICRP 23(2).

Kreuzer, M., Brachner, A., Lehmann, F., 2002. Characteristics of the German uranium miners cohort study. Health Phys. 83, 26–34.

Kreuzer, M., Walsh, L., Schnelzer, M., et al., 2008. Radon and risk of extrapulmonary cancers: results of the German uranium miners' cohort study, 1960–2003. Br. J. Cancer 99, 1946–1953.

Kreuzer, M., Schnelzer, M., Tschense, A., et al., 2010. Cohort profile: the German uranium miners cohort study (WISMUT cohort), 1946–2003. Int. J. Epidemiol. 39, 980–987.

Kusiak, R.A., Ritchie, A.C., Muller, J., Springer, J., 1993. Mortality from lung cancer in Ontario uranium miners. Br. J. Ind. Med. 50, 920–928.

L'Abbé, K.A., Howe, G.R., Burch, J.D., et al., 1991. Radon exposure, cigarette smoking, and other mining experience in the Beaverlodge uranium miners cohort. Health Phys. 60, 489–495.

Lane, R.S., Frost, S.E., Howe, G.R., et al., 2010. Mortality (1950–1999) and cancer incidence (1969–1999) in the cohort of Eldorado uranium workers. Radiat. Res. 174, 773–785.

Laurier, D., Tirmarche, M., Mitton, N., et al., 2004. An update of cancer mortality among the French cohort of uranium miners: extended follow-up and new source of data for causes of death. Eur. J. Epidemiol. 19, 139–146.

Leuraud, K., Billon, S., Bergot, D., et al., 2007. Lung cancer risk associated to exposure to radon and smoking in a case–control study of French uranium miners. Health Phys. 92, 371–378.

Lubin, J.H., Qiao, Y.-L., Taylor, P.R., et al., 1990. Quantitative evaluation of the radon and lung cancer association in a case control study of Chinese tin miners. Cancer Res. 50, 174–180.

Lubin, J., Boice, J.D., Edling, J.C., et al., 1994. Radon and Lung Cancer Risk: A Joint Analysis of 11 Underground Miner Studies. Publication No. 94-3644. US National Institutes of Health, Bethesda, MD.

Lubin, J.H., Boice Jr., J.D., Edling, C., et al., 1995. Radon-exposed underground miners and inverse dose-rate (protraction enhancement) effects. Health Phys. 69, 494–500.

Lubin, J.H., Tomášek, L., Edling, C., et al., 1997. Estimating lung cancer mortality from residential radon using data for low exposures of miners. Radiat. Res. 147, 126–134.

NRC, 1999. Health Effects of Exposure to Radon. BEIR VI Report. National Academy Press, Washington, DC.

Qiao, Y.L., Taylor, P.R., Yao, S.X., et al., 1989. Relation of radon exposure and tobacco use to lung cancer among tin miners in Yunnan Province, China. Am. J. Ind. Med. 16, 511–521.

Rogel, A., Laurier, D., Tirmarche, M., Quesne, B., 2002. Lung cancer risk in the French cohort of uranium miners. J. Radiol. Prot. 22, A101–A106.

Schnelzer, M., Hammer, G.P., Kreuzer, M., et al., 2010. Accounting for smoking in the radon-related lung cancer risk among German uranium miners: results of a nested case–control study. Health Phys. 98, 20–28.

Schubauer-Berigan, M.K., Daniels, R.D., Pinkerton, L.E., 2009. Radon exposure and mortality among white and American Indian uranium miners: an update of the Colorado Plateau cohort. Am. J. Epidemiol. 169, 718–730.

Thomas, D., Pogoda, J., Langholz, B., Mack, W., 1994. Temporal modifiers of the radon-smoking interaction. Health Phys. 66, 257–262.

Tirmarche, M., Laurier, D., Bergot, D., et al., 2003. Quantification of Lung Cancer Risk After Low Radon Exposure and Low Exposure Rate: Synthesis from Epidemiological and Experimental Data. Final Scientific Report, February 2000–July 2003. Contract FIGH-CT1999-0013. European Commission DG XI, Brussels.

Tirmarche, M., Laurier, D., Bochicchio, F., et al., 2010. Final Scientific Report of Alpha Risk Project. Funded by the European Commission EC FP6 (Ref. FI6R-CT-2005-516483). European Commission DG XII, Brussels. <http://www.alpha-risk.org>.

Tomášek, L., 2002. Czech miner studies of lung cancer risk from radon. J. Radiol. Prot. 22, A107–A112.

Tomášek, L., Placek, V., 1999. Radon exposure and lung cancer risk: Czech cohort study. Radiat. Res. 152, S59–S63.

Tomášek, L., Zarska, H., 2004. Lung cancer risk among Czech tin and uranium miners – comparison of lifetime detriment. Neoplasma 51, 255–260.

Tomášek, L., Plaček, V., Müller, T., et al., 2003. Czech studies of lung cancer risk from radon. Int. J. Low Radiat. 1, 50–62.

Tomášek, L., Rogel, A., Tirmarche, M., et al., 2008. Lung cancer in French and Czech uranium miners – risk at low exposure rates and modifying effects of time since exposure and age at exposure. Radiat. Res. 169, 125–137.

UNSCEAR, 2009. UNSCEAR 2006 Report, Annex E. Sources-to-Effects Assessment for Radon in Homes and Workplaces. United Nations, New York.

Vacquier, B., Caer, S., Rogel, A., 2008. Mortality risk in the French cohort of uranium miners: extended follow-up 1946–1999. Occup. Environ. Med. 65, 597–604.

Vacquier, B., Rogel, A., Leuraud, K., et al., 2009. Radon-associated lung cancer risk among French uranium miners: modifying factors of the exposure-risk relationship. Radiat. Environ. Biophys. 48, 1–9.

Villeneuve, P.J., Morrison, H.I., Lane, R., 2007. Radon and lung cancer risk: an extension of the mortality follow-up of the Newfoundland fluorspar cohort. Health Phys. 92, 157–169.

Walsh, L., Dufey, F., Tschense, A., et al., 2010a. Radon and the risk of cancer mortality–internal Poisson models for the German uranium miners cohort. Health Phys. 99, 292–300.

Walsh, L., Tschense, A., Schnelzer, M., et al., 2010b. The influence of radon exposures on lung cancer mortality in German uranium miners, 1946–2003. Radiat. Res. 173, 79–90.

Yao, S.X., Lubin, J.H., Qiao, Y.L., et al., 1994. Exposure to radon progeny, tobacco use and lung cancer in a case–control study in southern China. Radiat. Res. 138, 326–336.

4. ASSESSMENT OF THE DETRIMENT FROM EXPOSURE TO RADON AND ITS PROGENY

4.1. Risks other than lung cancer

(36) Radon and its progeny deliver substantially more dose to the lungs than systemic organs and the gastrointestinal tract. Nevertheless, calculations indicate that small doses may be received by the red bone marrow and other systemic organs (Khursheed, 2000; Kendall and Smith, 2002, 2005; Marsh et al., 2008).

(37) Studies of underground miners have not generally shown any excess of cancer other than lung cancer to be associated with radon exposure (Darby et al., 1995; NRC, 1999; UNSCEAR, 2009). Some associations have been suggested in individual studies, but they have not been replicated in other studies and no consistent pattern has emerged. For example, recent studies in the Czech Republic indicated an association with the incidence of chronic lymphocytic leukaemia (Rericha et al., 2006), but this finding was not confirmed by other studies in the Czech Republic (Tomášek and Malatova, 2006) and Germany (Möhner et al., 2006, 2010). Also, an excess of larynx cancer suggested in some analyses was not confirmed in other studies (Laurier et al., 2004; Möhner et al., 2008). Recent studies noted specific excesses or trends with radon exposure for non-Hodgkin's lymphoma; multiple myeloma; and kidney, liver, and stomach cancers (Vacquier et al., 2008; Kreuzer et al., 2008; Schubauer-Berigan et al., 2009). However, such observations have not been confirmed by other studies.

(38) Epidemiological studies have been conducted on the possible association between leukaemia and indoor radon concentrations (Laurier et al., 2001; Raaschou-Nielsen, 2008). An association between childhood leukaemia and domestic radon exposure has been observed in some ecological studies, including the recent findings of Evrard et al. (2005, 2006). Several large-scale case–control studies that included alpha track measurements in the homes of all subjects were unable to confirm an association between radon exposure and the risk of leukaemia (Lubin et al., 1998; Steinbuch et al., 1999; UK Childhood Cancer Study Investigators, 2002). A recent study in Denmark suggested a significant positive association between radon concentration, estimated on the basis of comprehensive modelling, and acute lymphocytic leukaemia, whereas a non-significant negative association was observed for acute non-lymphocytic leukaemia (Raaschou-Nielsen et al., 2008). A recent review concluded that an association between indoor exposure to radon and childhood leukaemia might exist, but the current epidemiological evidence is weak and further research with better designed studies is needed (Raaschou-Nielsen, 2008).

(39) In conclusion, the review of the available epidemiological evidence shows no consistent evidence for an association between radon concentration and cancer, other than lung cancer.

(40) It is noted that most available data relate to adult populations. While dosimetric calculations indicate that doses per unit exposure should not differ appreciably between children and adults (see Annex B, Para. B10), more information is needed to quantify the effects of exposures received during childhood.

4.2. Calculation of lung cancer lifetime risk estimates for underground miners

(41) Most miner studies have demonstrated the existence of time-modifying factors of the relationship between cumulated radon exposure and risk of lung cancer, such as age at exposure, or TSE. Due to variations in the characteristics of the study populations (attained age, duration of follow-up), the direct comparison of ERRs from different cohorts may be misleading. Such variations can be taken into account in calculation of the lifetime risk associated with a specific exposure scenario (Thomas et al., 1992). Calculation of lifetime risk requires:

- risk coefficients derived from an epidemiological study or studies, with or without modifying factors such as attained age;
- a projection model, enabling extrapolation of risk outside the range considered by the epidemiological study (exposure range, sex, age) and transport to other populations;
- background reference rates for all-cause and lung cancer mortality; and a scenario of exposure to radon concentrations.

(42) This approach was used in *Publication 65* to estimate the risk of lung cancer associated with prolonged exposure to radon concentrations, based upon studies of underground miners (ICRP, 1993). Several lifetime risk estimates have been published subsequently (NRC, 1999; EPA, 2003; Tomášek et al., 2008a), but these cannot be compared easily due to differences in the nature of the estimates or the underlying assumptions. In the present report, the focus is on estimates of the lifetime excess absolute risk (LEAR) of lung cancer mortality from radon and its progeny, as considered in *Publication 65* (ICRP, 1993), and the estimates derived for background rates corresponding to a specific country are excluded. Priority is given to models derived from pooled analyses rather than from single studies. The published estimates are summarised in Table 4.1.

(43) The exposure scenario considered in estimating the LEAR shown in Table 4.1 is the same as that proposed in *Publication 65* (ICRP, 1993): constant low-level exposure to 2 WLM per year during adulthood from 18 to 64 years of age, with risk estimated up to 90 or 94 years of age. Using the reference background rates of lung cancer from *Publication 60* (ICRP, 1991), *Publication 65* (ICRP, 1993) adopted a LEAR for lung cancer (also denoted as the nominal probability coefficient or fatality probability) of 2.8×10^{-4} per WLM for radon exposure. Since the detriment was entirely due to lung cancer mortality, the Commission adopted a total detriment coefficient equal to this fatality coefficient (ICRP, 1993).

(44) Applying the same risk coefficient as in *Publication 65* (ICRP, 1993) to the reference background rates found in *Publication 103* (ICRP, 2007), Tomášek et al. (2008b) calculated a LEAR of lung cancer of 2.7×10^{-4} per WLM. This comparison shows that the modification of the reference population for background cancer rates between *Publication 60* and *Publication 103* has only a small impact on the estimated LEAR.

(45) Using the same exposure scenario as in *Publication 65* (ICRP, 1993) and reference background rates from *Publication 103* (ICRP, 2007), Tomášek et al.

Table 4.1. Estimates of the lifetime excess absolute risk (LEAR) of lung cancer associated with concentrations of radon and its progeny in underground mines [*Publication 65* scenario of constant exposure to 2 working level months (WLM) per year from 18 to 64 years of age].

Primary risk model	Projection model	Background reference rates	LEAR $(10^{-4}$ per WLM)	Reference
Publication 65 (ICRP, 1993)	Relative risk	*Publication 60* (ICRP, 1991)	2.8	ICRP (1993)
Publication 65 (ICRP, 1993)	Relative risk	*Publication 103* (ICRP, 2007)	2.7	Tomášek et al. (2008b)
BEIR VI model TSE–age–concentration (NRC, 1999)	Relative risk	*Publication 103* (ICRP, 2007)	5.3	Tomášek et al. (2008b)
Czech–French joint model* (Tomášek et al., 2008a)	Relative risk	*Publication 103* (ICRP, 2007)	4.4	Tomášek et al. (2008b)

Publication 60 reference rates: averaged over males and females and over five countries.
Publication 103 reference rates: averaged over males and females and over Asian and Euro-American populations.
* Model relying on periods of work with the best quality of exposure assessment.

(2008b) also calculated the LEAR using the BEIR VI TSE–age–concentration model (NRC, 1999). This model relies on the combined analysis of data from 11 cohorts of miners, and takes into account the modifying effects of attained age, TSE, and exposure rate (note that the scenario only corresponds to the lowest category of exposure rate). The LEAR estimate based on this model was 5.3×10^{-4} per WLM.

(46) Based on the same assumptions [exposure scenario from *Publication 65* (ICRP, 1993) and reference background rates from *Publication 103* (ICRP, 2007)], Tomášek et al. (2008b) calculated the LEAR using the model developed from the combined analysis of the Czech–French cohorts (Tomášek et al., 2008a). This model used exposure data for the periods of work with the best quality of exposure assessment. It took account of the modifying effects of age at exposure and TSE. As the analysis focused on miners with low levels of exposure, no effect of exposure rate was observed in this analysis (Tomášek et al., 2008a). The LEAR estimate based on the Czech–French model was 4.4×10^{-4} per WLM (Tomášek et al., 2008a).

(47) Table 4.1 shows a substantial increase in the LEAR estimated using both the BEIR VI model and the Czech–French model compared with the LEAR estimated using the model from *Publication 65* (ICRP, 1993). Other published lifetime estimates, based on specific national rates and therefore not directly comparable with the LEAR estimated in *Publication 65* (ICRP, 1993), also support a tendency for an increase in the estimated lifetime risks compared with earlier values (EPA, 2003). This increase in LEAR estimates is related, in part, to consideration of chronic low-rate exposures and, in part, to the increase in the estimated ERR per WLM observed in recent studies.

(48) Additional LEAR calculations were performed by the Task Group in order to validate the published results, and to provide a sensitivity analysis of the different underlying hypotheses using different models, scenarios, and background rates. Some calculations were performed independently by different experts to provide an internal quality check. Results confirmed the higher LEAR estimated using the

BEIR VI model and the Czech–French model. In addition to these models derived from pooled analyses, other recent models obtained from single studies were also considered [French CEA-AREVA cohort (Vacquier et al., 2008), Canadian Eldorado cohort (Howe, 2006), German Wismut cohort (Grosche et al., 2006)]. These studies show that the estimated LEAR can vary from about 3 to 7×10^{-4} per WLM according to the model used. They also illustrate the sensitivity of the estimate to the choice of the model, and re-inforced the preference for models derived from pooled analyses. Other calculations also illustrated the sensitivity of LEAR estimates to background rates. Using the rates for Euro-American males instead of the reference rates averaged over males and females and over Euro-American and Asian populations (ICRP, 2007), the estimated LEAR is about 7×10^{-4} per WLM. This difference is due to the higher background rate of lung cancer among Euro-American males. Conversely, using lower background rates of lung cancer (such as females or non-smokers) would lead to a lower estimated LEAR per WLM.

(49) Based on the previous considerations, the Commission now recommends that a LEAR of 5×10^{-4} per WLM [14×10^{-5} per (mJh/m^3)] should be used as the nominal probability coefficient for radon- and radon-progeny-induced lung cancer, replacing the *Publication 65* value of 2.8×10^{-4} per WLM [8×10^{-5} per (mJh/m^3)]. Current knowledge of radon-associated risks for organs other than the lungs does not justify the selection of a detriment coefficient different from the fatality coefficient for radon exposure. The estimated LEAR of lung cancer mortality, corresponding to the attributable probability of fatal lung cancer (or nominal fatality probability coefficient), is therefore considered to reflect the lifetime detriment associated with exposure to radon and its progeny.

4.3. Comparison of results from underground mine and domestic exposures

(50) The comparison of results obtained from miner studies and indoor studies is not straightforward. This is mainly due to the use of different epidemiological designs (mostly cohort studies for miners and case–control studies for indoor exposures) as well as different measures of exposure (WLM in mines, radon gas concentrations in homes). The miner studies have the advantage of considering the distribution over time of the individual radon exposures and therefore enable consideration of the modifying effects of age and TSE, but are often unable to consider the effect of cofactors, such as smoking. The domestic case–control studies have the advantage of providing detailed information about many potential cofactors, but contemporary measures must be used to estimate prior radon concentrations during previous decades. They generally only consider the average radon concentration in a home over a given period, and are not able to analyse potential time modifiers of the exposure–risk relationship.

(51) Estimated primary risk coefficients are presented in Tables 2.1 and 2.2 for indoor studies, and Tables 3.1 and 3.2 (and Annex A) for miner studies. According to *Publication 65*, assuming an occupancy of 7000 h/year and F = 0.4, a concentration of 1 Bq/m^3 radon gas leads to indoor exposure of 4.40×10^{-3} WLM (ICRP, 1993). Most indoor case–control studies have estimated radon concentrations for periods of

30 or 35 years before diagnosis, with an exposure lag time of 5 years. Therefore, considering a period of 30 years (i.e. the last 35 years before diagnosis with a lag time of 5 years) and a time-weighted averaged concentration of 100 Bq/m^3, the cumulative exposure of 2.1×10^7 hBq/m^3 corresponds to a cumulative exposure of approximately 13 WLM assuming F = 0.4. Using these values, an ERR per 100 Bq/m^3 of 0.16 for indoor exposures (as obtained in the European pooling study with uncertainty correction; Darby et al., 2006) corresponds to an ERR of 1.2 per 100 WLM, which is similar to the value obtained in the BEIR VI analysis restricted to low levels of exposure below 50 WLM (NRC, 1999; see Table 3.2). This approach indicates reasonably good agreement between the risk coefficients estimated for lung cancer mortality from indoor studies and miner studies at low levels of exposure. The same reasoning has been presented by several authors and led to the same conclusion (Zielinski et al., 2006; Tomášek et al., 2008a; UNSCEAR, 2009).

(52) The above approach does not consider the modifying effects of age and TSE on the exposure–risk relationship demonstrated by miner studies. Lifetime estimates of the risk of lung cancer can account for these modifying factors, and provide another method for comparing the results of miner studies with those of indoor radon investigations. Nevertheless, due to differences in background rates, duration of life considered, and exposure scenarios, considerable caution is needed in comparing published lifetime estimates obtained from miner studies (ICRP, 1993; NRC, 1999; EPA, 2003; Tomášek et al., 2008a) and indoor studies (Darby et al., 2006).

(53) To enable comparison of estimated risks between miner studies and the European indoor study, additional calculations were performed using parameters chosen to respect the characteristics of the available data as closely as possible. A specific scenario was elaborated in order to reflect the characteristics of the individuals included in the European indoor study (attained age of 70 years corresponding to the average age at diagnosis, constant exposure to 100 Bq/m^3 over a time window of 5–30 years before diagnosis). To reflect the fact that miner studies provide risk estimates for males, the ERR per 100 Bq/m^3 of 0.25 obtained in the European pooled study for males was used (Darby et al., 2006). Using these parameters, the values for cumulated absolute risk up to 70 years of age estimated for two pooled analyses of miner studies (BEIR VI and French–Czech) and for the European pooled analysis of indoor exposures were 3.5, 2.7, and 2.7×10^{-4} per WLM, respectively.

(54) In conclusion, the currently available results show reasonably good consistency between lung cancer risk estimates obtained from miner studies and indoor studies.

4.4. References

Darby, S.C., Whitley, E., Howe, G.R., et al., 1995. Radon and cancers other than lung cancer in underground miners: a collaborative analysis of 11 studies. J. Natl. Cancer Inst. 87, 378–384.

Darby, S., Hill, D., Deo, H., et al., 2006. Residential radon and lung cancer – detailed results of a collaborative analysis of individual data on 7148 persons with lung cancer and 14,208 persons without lung cancer from 13 epidemiological studies in Europe. Scand. J. Work Environ. Health 32 (Suppl. 1), 1–84.

EPA, 2003. Assessment of Risks from Radon in Homes. Publication EPA 402-R-03-003. Office of Air and Radiation, United States Environmental Protection Agency, Washington, DC.

Evrard, A.S., Hémon, D., Billon, S., et al., 2005. Childhood leukemia incidence and exposure to indoor radon, terrestrial and cosmic gamma radiation. Health Phys. 90, 569–579.

Evrard, A.S., Hémon, D., Morin, A., et al., 2006. Childhood leukaemia incidence around French nuclear installations using a geographic zoning based on gaseous release dose estimates. Br. J. Cancer 94, 1342–1347.

Grosche, B., Kreuzer, M., Kreisheimer, M.A., 2006. Lung cancer risk among German male uranium miners: a cohort study, 1946–1998. Br. J. Cancer 95, 1280–1287.

Howe, G.R., 2006. Updated Analysis of the Eldorado Uranium Miner's Cohort: Part I of the Saskatchewan Uranium Miner's Cohort Study. RSP-0205. Columbia University, New York.

ICRP, 1991. 1990 Recommendations of the International Commission on Radiological Protection. ICRP Publication 60. Ann. ICRP 21(1–3).

ICRP, 1993. Protection against radon-222 at home and at work. ICRP Publication 65. Ann. ICRP 23(2).

ICRP, 2007. The 2007 Recommendations of the International Commission on Radiological Protection. ICRP Publication 103. Ann. ICRP 37(2–4).

Kendall, G.M., Smith, T.J., 2002. Doses to organs and tissues from radon and its decay products. J. Radiol. Prot. 22, 389–406.

Kendall, G.M., Smith, T.J., 2005. Doses from radon and its decay products to children. J. Radiol. Prot. 25, 241–256.

Khursheed, A., 2000. Doses to systemic tissue from radon gas. Radiat. Prot. Dosim. 88, 171–181.

Kreuzer, M., Walsh, L., Schnelzer, M., et al., 2008. Radon and risk of extrapulmonary cancers: results of the German uranium miners' cohort study, 1960–2003. Br. J. Cancer 99, 1946–1953.

Laurier, D., Valenty, M., Tirmarche, M., 2001. Radon exposure and the risk of leukemia: a review of epidemiological studies. Health Phys. 81, 272–288.

Laurier, D., Tirmarche, M., Mitton, N., et al., 2004. An update of cancer mortality among the French cohort of uranium miners: extended follow-up and new source of data for causes of death. Eur. J. Epidemiol. 19, 139–146.

Lubin, J.H., Linet, M.S., Boice Jr., J.D., et al., 1998. Case–control study of childhood acute lymphoblastic leukemia and residential radon exposure. J. Natl. Cancer Inst. 90, 294–300.

Marsh, J.W., Bessa, Y., Birchall, A., et al., 2008. Dosimetric models used in the Alpha-Risk project to quantify exposure of uranium miners to radon gas and its progeny. Radiat. Prot. Dosim. 130, 101–106.

Möhner, M., Lindtner, M., Otten, H., Gille, H.G., 2006. Leukemia and exposure to ionizing radiation among German uranium miners. Am. J. Ind. Med. 49, 238–248.

Möhner, M., Lindtner, M., Otten, H., 2008. Ionizing radiation and risk of laryngeal cancer among German uranium miners. Health Phys. 95, 725–733.

Möhner, M., Gellissen, J., Marsh, J.W., et al., 2010. Occupational and diagnostic exposure to ionizing radiation and leukemia risk among German uranium miners. Health Phys. 99, 314–321.

NRC, 1999. Health Effects of Exposure to Radon. BEIR VI Report. National Academy Press, Washington, DC.

Raaschou-Nielsen, O., 2008. Indoor radon and childhood leukaemia. Radiat. Prot. Dosim. 132, 175–181.

Raaschou-Nielsen, O., Andersen, C.E., Andersen, H.P., et al., 2008. Domestic radon and childhood cancer in Denmark. Epidemiology 19, 536–543.

Rericha, V., Kulich, M., Rericha, R., et al., 2006. Incidence of leukemia, lymphoma, and multiple myeloma in Czech uranium miners: a case–cohort study. Environ. Health Perspect. 114, 818–822.

Schubauer-Berigan, M.K., Daniels, R.D., Pinkerton, L.E., 2009. Radon exposure and mortality among white and American Indian uranium miners: an update of the Colorado Plateau cohort. Am. J. Epidemiol. 169, 718–730.

Steinbuch, M., Weinberg, C.R., Buckley, J.D., et al., 1999. Indoor residential radon exposure and risk of childhood acute myeloid leukaemia. Br. J. Cancer 81, 900–906.

Thomas, D., Darby, S., Fagnani, F., et al., 1992. Definition and estimation of lifetime detriment from radiation exposures: principles and methods. Health Phys. 63, 259–272.

Tomášek, L., Malatova, I., 2006. Leukaemia and lymphoma among Czech uranium miners. Med. Radiat. Radiat. Saf. 51, 74–79.

Tomášek, L., Rogel, A., Tirmarche, M., et al., 2008a. Lung cancer in French and Czech uranium miners – risk at low exposure rates and modifying effects of time since exposure and age at exposure. Radiat. Res. 169, 125–137.

Tomášek, L., Rogel, A., Tirmarche, M., et al., 2008b. Dose conversion of radon exposure according to new epidemiological findings. Radiat. Prot. Dosim. 130, 98–100.

UK Childhood Cancer Study Investigators, 2002. The United Kingdom Childhood Cancer Study of exposure to domestic sources of ionising radiation: 1: radon gas. Br. J. Cancer 86, 1721–1726.

UNSCEAR, 2009. UNSCEAR 2006 Report, Annex E. Sources-to-Effects Assessment for Radon in Homes and Workplaces. United Nations, New York.

Vacquier, B., Caer, S., Rogel, A., 2008. Mortality risk in the French cohort of uranium miners: extended follow-up 1946–1999. Occup. Environ. Med. 65, 597–604.

Zielinski, J.M., Carr, Z., Repacholi, M., Krewski, D., 2006. World Health Organization's International Radon Project. J. Toxicol. Environ. Health A 69, 759–769.

5. CONCLUSIONS

(55) The present review and analysis of the epidemiology of radon leads to the following conclusions.

- There is compelling evidence from cohort studies of underground miners and from case–control studies of residential radon exposures that radon and its progeny can cause lung cancer. For solid tumours other than lung cancer, and also for leukaemia, there is currently no convincing or consistent evidence of any excesses associated with exposure to radon and its progeny.
- The three pooled residential case–control studies in Europe, North America, and China gave similar results and showed that the risk of lung cancer increases by at least 8% for an increase in radon concentration of 100 Bq/m^3 (Lubin et al., 2004; Darby et al., 2005; Krewski et al., 2006).
- After correcting for random uncertainties in the radon activity concentration measurements, the European pooled residential case–control study gave an ERR of 16% (95% CI: 5–31%) per 100 Bq/m^3 increase (Darby et al., 2005). This value may be considered to be a reasonable estimate for risk management purposes at relatively low and prolonged radon exposures in homes, considering that this risk is linked to an exposure period of at least 25 years.
- There is evidence from the European pooled residential case–control study that there is a risk of lung cancer even at levels of long-term average radon concentration below 200 Bq/m^3 (Darby et al., 2005).
- The cumulative risk of lung cancer up to 75 years of age for lifelong non-smokers is estimated to be 0.4%, 0.5%, and 0.7% for radon activity concentrations of 0, 100, and 400 Bq/m^3, respectively. The cumulative risks of lung cancer for lifelong smokers by 75 years of age are close to 10%, 12%, and 16% for radon activity concentrations of 0, 100, and 400 Bq/m^3, respectively (Darby et al., 2005, 2006). Cigarette smoking remains the most important cause of lung cancer.
- Appropriate comparisons of lung cancer risk estimates from miner studies and indoor studies show good consistency.
- Based upon a review of epidemiological studies of underground miners, including studies with relatively low levels of exposure, a detriment-adjusted nominal risk coefficient of 5×10^{-4} per WLM (0.14 per Jh/m^3) is adopted for the lung detriment per unit exposure to radon. This value of 5×10^{-4} per WLM (0.14 per Jh/m^3) is derived from recent studies considering exposure during adulthood, and is close to twice the value calculated in *Publication 65* (ICRP, 1993).

(56) Risk estimates obtained from indoor epidemiological studies are sufficiently robust to enable protection of the public to be based on residential concentration levels. *Publication 65* (ICRP, 1993) recommended that doses from radon and its progeny should be calculated using a dose conversion convention based on miner epidemiological studies. No such conversion convention is proposed in the present report.

(57) For occupational protection purposes, dose estimates are required to demonstrate compliance with limits and constraints. In addition to the review of epidemi-

ological data, published dose calculations for radon and progeny were also reviewed (see Annex B). Published values of effective dose from inhalation of radon progeny derived using the Human Respiratory Tract Model (HRTM) range from about 10 to 20 mSv per WLM [3–6 mSv per (mJh/m^3)] depending on the exposure scenario. It should be noted that these coefficients are larger by a factor of two or more than the conversion coefficients derived from *Publication 65* (ICRP, 1993).

(58) The Commission now proposes to treat radon and its progeny in the same way as other radionuclides within the system of protection and to publish dose coefficients (dose per unit exposure) in the near future. Doses from radon and its progeny will be calculated using ICRP biokinetic and dosimetric models, including the HRTM in *Publication 66* (ICRP, 1994) and ICRP systemic models. This will apply to thoron and its progeny, as well as radon and its progeny (see Annex B). Reference ICRP dose coefficients per unit exposure to radon and its progeny will be published for different reference conditions of exposure, with specified aerosol characteristics and F values.

5.1. References

Darby, S., Hill, D., Auvinen, A., et al., 2005. Radon in homes and risk of lung cancer: collaborative analysis of individual data from 13 European case–control studies. Br. Med. J. 330, 223–227.

Darby, S., Hill, D., Deo, H., et al., 2006. Residential radon and lung cancer – detailed results of a collaborative analysis of individual data on 7148 persons with lung cancer and 14,208 persons without lung cancer from 13 epidemiological studies in Europe. Scand. J. Work Environ. Health 32 (Suppl. 1), 1–84.

ICRP, 1993. Protection against radon-222 at home and at work. ICRP Publication 65. Ann. ICRP 23(2).

ICRP, 1994. Human respiratory tract model for radiological protection. ICRP Publication 66. Ann. ICRP 24(1–3).

Krewski, D., Lubin, J.H., Zielinski, J.M., et al., 2006. A combined analysis of North American case–control studies of residential radon and lung cancer. J. Toxicol. Environ. Health Part A 69, 533–597.

Lubin, J.H., Wang, Z.Y., Boice Jr., J.D., et al., 2004. Risk of lung cancer and residential radon in China: pooled results of two studies. Int. J. Cancer 109, 132–137.

ANNEX A. RESULTS FROM EPIDEMIOLOGICAL STUDIES OF UNDERGROUND MINERS

Table A.1. Characteristics of the cohorts used in *Publication 65* (ICRP, 1993).

Place	Country	Type of mine	Follow-up period	No. of miners	Cumu-lative exposure WLM	Person-years[a]	ERR per 100 WLM	95% CI
Colorado[a]	USA	Uranium	1951–1982	2975	510	66,237	0.60	0.30–1.42
Ontario	Canada	Uranium	1955–1981	11,076	37	217,810	1.42	0.60–3.33
New Mexico	USA	Uranium	1957–1985	3469	111	66,500	1.81	0.71–5.46
Beaverlodge	Canada	Uranium	1950–1980	6895	44	114,170	1.31	0.60–3.01
West Bohemia	Czech Republic	Uranium	1953–1985	4042	227	97,913	1.70	1.21–2.41
CEA-COGEMA	France	Uranium	1946–1985	1785	70	44,005	0.60	0.00–1.63
Malmberget	Sweden	Iron	1951–1976	1292	98	27,397	1.42	0.30–9.57
Total				**31,486**	**120**	**635,022**	**1.34**	**0.82–2.13**

WLM, working level month; ERR, excess relative risk; CI, confidence interval.

[a] Less than 2000 WLM.

Table A.2. Characteristics of the cohorts considered in the BEIR VI report (NRC, 1999).

Place	Country	Type of mine	Follow-up period	No. of miners	Cumulative exposure WLM	Person-years[a]	ERR per 100 WLM	SE
Yunnan	China	Tin	1976–1987	13,649	286.0	134,842	0.17	
W-Bohemia	Czech Republic	Uranium	1952–1990	4320	196.8	102,650	0.67	
Colorado	USA	Uranium	1950–1990	3347	578.6	79,556	0.44	
Ontario	Canada	Uranium	1955–1986	21,346	31.0	300,608	0.82	
Newfoundland	Canada	Fluorspar	1950–1984	1751	388.4	33,795	0.82	
Malmberget	Sweden	Iron	1951–1991	1294	80.6	32,452	1.04	
New Mexico	USA	Uranium	1943–1985	3457	110.9	46,800	1.58	
Beaverlodge	Canada	Uranium	1950–1980	6895	21.2	67,080	2.33	
Port Radium	Canada	Uranium	1950–1980	1420	243.0	31,454	0.24	
Radium Hill	Australia	Uranium	1948–1987	1457	7.6	24,138	2.75	
CEA-COGEMA	France	Uranium	1948–1986	1769	59.4	39,172	0.51	
Total				**60,606**	**164.4**	**888,906**	**0.59**	**1.32**

WLM, working level month; ERR, excess relative risk; SE, multiplicative standard error.

[a] Among exposed.

Table A.3. Characteristics of the cohorts considered by UNSCEAR (2009).

Place	Country	Type of mine	Follow-up period	No. of miners	Cumu-lative exposure WLM	Person-years	ERR per 100 WLM	95% CI
Colorado	USA	Uranium	1950–1990	3347	807	75,032	0.42	0.3–0.7
Newfoundland	Canada	Fluorspar	1951–2001	1742	378	70,894	0.47	0.28–0.65
Yunnan	China	Tin	1976–1987	13,649	277	135,357	0.16	0.1–0.2
Wismut	Germany	Uranium	1946–1998	59,001	242	1,801,626	0.21	0.18–0.24
Malmberget	Sweden	Iron	1951–1990	1415	81	32,452	0.95	0.1–4.1
West Bohemia	Czech Republic	Uranium	1952–1999	9979	70	261,428	1.60	1.2–2.2
CEA-COGEMA	France	Uranium	1946–1994	5098	37	133,521	0.80	0.3–1.4
Ontario	Canada	Uranium	1955–1986	21,346	31	319,701	0.89	0.5–1.5
Beaverlodge	Canada	Uranium	1950–1999	10,050	23	285,964	0.96	0.56–1.56
Total				**125,627**		**3,115,975**	**0.59**	**0.35–1.0**

WLM, working level month; ERR, excess relative risk; CI, confidence interval.

A.1. References

ICRP, 1993. Protection against radon-222 at home and at work. ICRP Publication 65. Ann. ICRP 23(2).

NRC, 1999. Health Effects of Exposure to Radon. BEIR VI Report. National Academy Press, Washington, DC.

UNSCEAR, 2009. UNSCEAR 2006 Report, Annex E. Sources-to-Effects Assessment for Radon in Homes and Workplaces. United Nations, New York.

ANNEX B. DOSIMETRY

B.1. Radon

(B1) The equivalent dose to the lungs following the inhalation of radon and its short-lived progeny can be calculated using the HRTM (ICRP, 1994) and other models of the human respiratory tract. Nearly the entire lung dose arises from inhalation of radon progeny and not from radon itself, as almost all of the gas that is inhaled is subsequently exhaled. However, a large proportion of the inhaled radon progeny deposits in the respiratory airways of the lung. Due to their short half-lives (<30 min), dose is delivered to the lung tissues before clearance can take place, either by absorption into blood or by particle transport to the alimentary tract. Two of the short-lived radon progeny (polonium-218 and polonium-214) decay by alpha particle emission, and it is the energy from these alpha particles that accounts for the relatively high dose to the lung. In comparison, doses to systemic organs and gastrointestinal tract regions are low; effective dose is dominated by the equivalent dose to the lungs.

(B2) The radon progeny aerosol in the atmosphere is created in two steps. After decay of the radon gas, the freshly formed radionuclides (polonium-218, lead-214, and bismuth-214) react rapidly (<1 s) with trace gases and vapours, and grow by cluster formation to form particles around 1 nm in size. These are referred to as 'unattached particles'. The unattached particles may also attach to existing aerosol particles in the atmosphere within 1–100 s, forming the so-called 'attached particles'. The attached particles can have a trimodal activity size distribution which can be described by a sum of three lognormal distributions (Porstendörfer, 2001). These comprise the nucleation mode with an activity median aerodynamic diameter (AMAD) between 10 nm and 100 nm, the accumulation mode with an AMAD of 100–400 nm, and a coarse mode with an AMAD >1 μm. Generally, the greatest activity fraction is in the accumulation mode, which has a geometric standard deviation of about 2.

(B3) A dosimetric model for the respiratory tract needs to describe the morphometry, the deposition of the inhaled material, clearance from the respiratory tract, and the location of target tissues and cells at risk. For radon progeny, it is the dose to the target cells in the bronchial and bronchiolar regions of the lung that are of importance. In comparison, the dose to the alveolar region is significantly lower (UNSCEAR, 1982; Marsh and Birchall, 2000).

(B4) ICRP (1987) used values of dose per unit exposure to radon based on an NEA (1983) review of available dosimetric models (Hofmann et al., 1980; Jacobi and Eisfeld, 1980; Jacobi and Eisfeld, 1982; James et al., 1982; Harley and Pasternack, 1982). UNSCEAR reports (1982, 1988, 1993) used similar estimates of dose from radon inhalation, and the 2000 report retained a value of effective dose of 5.7 mSv per WLM [1.6 mSv per (mJh/m^3), i.e. 9 nSv per (Bqh/m^3) of equilibrium equivalent concentration (EEC) of radon] for indoor and outdoor exposures (Table B.1). In its 2000 report, UNSCEAR recognised that more recent calculations with new dosimetric models resulted in higher values of dose conversion factor. However, because of the lower values calculated using the dose conversion convention (ICRP, 1993),

53

Table B.1. Published values of effective dose to an adult male from the inhalation of radon and its progeny calculated using dosimetric models.

Publication	Model type	Exposure scenario	Effective dose (mSv per WLM)	Effective dose [mSv per (mJh/m^3)]
ICRP (1987)	NEA (1983)	Indoors	6.4	1.8
		Outdoors	8.9	2.5
UNSCEAR (2000)	NEA (1983)	Indoors and outdoors	5.7	1.6
Harley et al. (1996)		Indoors and mines	9.6[a]	2.7
Porstendörfer (2001)	Zock et al. (1996)	Home[b]	8	2.3
		Workplace	11.5	3.2
		Outdoor	10.6	3.0
Winkler-Heil and Hofmann (2002)	Deterministic airway generation model	Home	7.6	2.1
Winkler-Heil et al. (2007)	Deterministic airway generation model	Mine	8.3	2.3
	Stochastic airway generation model	Mine	8.9	2.5
	HRTM (ICRP, 1994)	Mine	11.8	3.3
Marsh and Birchall (2000)	HRTM (ICRP, 1994)	Home	15	4.2
James et al. (2004)	HRTM (ICRP, 1994)	Mine[c]	20.9	5.9
		Homeb	21.1	6.0
Marsh et al. (2005)	HRTM (ICRP, 1994)	Mine	12.5	3.5
		Homeb	12.9	3.6

WLM, working level month; HRTM, Human Respiratory Tract Model.

[a] An absorbed dose of 6 mGy per WLM [1.7 mGy per (mJh/m^3)] was calculated for the bronchial region. The effective dose per unit exposure was then calculated with a radiation-weighting factor for alpha particles of 20 and a tissue-weighting factor of 0.08 (= $2/3 \times 0.12$) for the bronchial and bronchiolar regions of the lung (ICRP, 1993).

[b] Home without cigarette smoke.

[c] No hygroscopic growth was assumed.

UNSCEAR concluded that the previous value of 9 nSv per (Bqh/m^3) of EEC was well within the range of possible dose conversion factors, and therefore should continue to be used in dose evaluations (UNSCEAR, 2000, 2009).

(B5) Table B.1 also shows values of effective dose per unit exposure to radon progeny [mSv per WLM or mSv per (mJh/m^3)] calculated using the HRTM from *Publication 66* (ICRP, 1994) and other models, including deterministic airway generation models (Harley et al., 1996; Porstendörfer, 2001; Winkler-Heil and Hofmann, 2002) and a stochastic airway generation model (Winkler-Heil et al., 2007). Results of selected recent calculations are given in Table B.1 and in Marsh et al. (2010). More comprehensive tabulations of values published between 1956 and 1998 are given by UNSCEAR (2000).

(B6) The main sources of variability and uncertainty in calculation of the equivalent dose to the lungs per unit exposure to radon progeny include:

- the activity size distribution of the radon progeny aerosol;
- the breathing rates;
- the model used to predict aerosol deposition in the respiratory tract;

- the absorption of the radon progeny from lungs to blood;
- the identification of target cells and their location within bronchial and bronchiolar epithelium;
- the relative sensitivity of different cell types to radiation; and
- the regional differences in the radiation sensitivity of the lung.

Marsh and Birchall (2000) performed a sensitivity analysis to identify those HRTM parameters that significantly affect the equivalent dose to the lungs (H_{lung}) per unit exposure to radon progeny under conditions found in houses. Other sensitivity analyses have been reported (NCRP, 1984; NRC, 1991; Zock et al., 1996; Tokonami et al., 2003), and UNSCEAR (1988) noted that equivalent dose may vary by a factor of about 3 according to the target cells considered.

(B7) Winkler-Heil et al., 2007 compared the results of the effective dose for radon progeny inhalation obtained using the HRTM, a deterministic airway generation model, and a stochastic airway generation model with the same input parameter values. Similar results were obtained ranging from 8.3 to 11.8 mSv per WLM [2.3–3.3 mSv per (mJh/m^3)] (Table B.1). The authors noted that one of the important issues affecting the comparison is the averaging procedure for the doses calculated in airway generation models.

(B8) Porstendörfer (2001) calculated doses from exposure to radon progeny for different exposure scenarios using an airway generation model developed by Zock et al. (1996). The effective dose calculated for 'normal' aerosol conditions in homes, workplaces, and outdoors ranged from 8.0 to 11.5 mSv per WLM [2.3–3.3 mSv per (mJh/m^3)] (Table B.1). However, in places with one dominant aerosol source producing a high particle concentration (e.g. cigarette smoking or combustion aerosols by diesel engines), the effective dose was calculated to be lower, ranging from 4.2 to 7.1 mSv per WLM [1.2–2.0 mSv per (mJh/m^3)]. The activity size distributions and unattached fractions assumed for these calculations were based on their measurements in indoor and outdoor air, and in the air at different workplaces in Germany.

(B9) Baias et al. (2010) calculated dose conversion factors (mSv per WLM) with a stochastic airway generation model for four different categories of smokers. Physiological and morphological changes to the lungs induced by smoking were accounted for using aerosol parameter values fixed for a mine atmosphere. Doses calculated for a light short-term smoker only differed by about 1% from values for a non-smoker (7.2 mSv per WLM). For the light long-term smoker and the heavy short-term smoker, the effective dose per WLM was calculated to decrease by more than 15% due to the thickening of the mucus layer. However, for the heavy long-term smoker, the effective dose per WLM was postulated to increase by about a factor of two compared with the non-smoker, primarily due to impaired mucociliary clearance, higher breathing frequency, and reduced lung volume due to obstructive lung diseases.

(B10) James et al. (2004) calculated effective doses from radon progeny for mines and homes using the HRTM. The activity size distributions given in the BEIR VI report (NRC, 1999) were assumed. The authors calculated a range of values for mines [18–21 mSv per WLM; 5.1–5.9 mSv per (mJh/m^3)] and homes [16–21 mSv per WLM; 4.5–5.9 mSv per (mJh/m^3)] depending upon whether or not the attached particles double in size in the respiratory tract due to hygroscopic growth, and depending upon the presence or absence of cigarette smoke in homes. These estimates are higher compared with other estimates (Table B.1), mainly because the activity size distributions assumed differed from those used by other investigators. Marsh et al. (2005), also using the HRTM and activity size distributions for mines and homes based upon measurements carried out in Europe, calculated values of about 13 mSv per WLM [3.7 mSv per (mJh/m^3)] for mines and homes (Table B.1).

(B11) Calculations performed with the HRTM showed that the equivalent dose to the lungs per unit exposure is relatively insensitive to age (NRC, 1999; Marsh and Birchall, 2000; Kendall and Smith, 2005; Marsh et al., 2005). For example, the lung dose for an adult compared with that of a child (>1 year) only differs by about 10%. The reason for this is that there are competing effects that tend to cancel out. Children have lower breathing rates which decrease the intake and lung doses, while this is partly compensated by the smaller mass of target tissue which increases the doses. Also, children have smaller airways that increase deposition by diffusion, but this is also compensated in part by smaller residence times that decrease deposition by diffusion.

(B12) The values of effective dose from the inhalation of radon progeny derived from the HRTM range from about 10 to 20 mSv per WLM [3–6 mSv per (mJh/m^3)], depending on the exposure scenario (Table B.1). For typical aerosol conditions in homes and mines, the effective dose is about 13 mSv per WLM [3.7 mSv per (mJh/m^3)] (Marsh et al., 2005). However, assuming the same aerosol conditions as for a home but with a breathing rate for a standard worker (1.2 per m^3/h), which may be appropriate for an indoor workplace, the effective dose increases from 13 mSv per WLM [3.7 mSv per (mJh/m^3)] to about 20 mSv per WLM [6 mSv per (mJh/m^3)].

(B13) The Commission has concluded that radon and its progeny should be treated in the same way as any other radionuclide within the system of protection. In other words, doses from radon and its progeny should be calculated using ICRP biokinetic and dosimetric models, including the HRTM and the ICRP systemic models. One of the advantages of this approach is that doses to organs other than the lungs can also be calculated. ICRP will provide dose coefficients per unit exposure to radon and its progeny for different reference conditions of domestic and occupational exposure, with specified equilibrium factors and aerosol characteristics.

B.2. Thoron

(B14) Thoron (radon-222) gas is a decay product of radium-224 and is part of the thorium-232 decay series. Thoron has a short half-live (56 s) and decays into a series of solid short-lived radioisotopes, including lead-212 which has a half-life of 10.6 h. Due to the short half-life of thoron, it is less able than radon to escape from the point where it is formed. As a consequence, building materials are the most usual source of indoor thoron exposure.

(B15) As for radon, doses from the inhalation of thoron and its progeny are dominated by alpha particle emissions from decay of the progeny (Jacobi and Eisfeld, 1980, 1982). Due to its very short half-life, the gas activity concentration of thoron can vary substantially across a room, and so it is not possible to use the concentration of thoron gas in dose evaluation. Therefore, for control purposes, the potential alpha energy concentration of thoron progeny should be determined for the estimation of thoron exposure. However, it is usually sufficient to control the intake of lead-212 for protection purposes because the potential alpha energy per unit activity inhaled is about 10 times higher for lead-212 than for other thoron progeny (ICRP, 1987).

(B16) UNSCEAR (2000) and the BEIR VI Committee (NRC, 1999) presented data for the ratio of potential alpha energy concentration arising from thoron progeny to that from radon progeny. The values ranged from 0.1 to 5. The highest values were for woodframe and mud houses found in Japan, and for some houses in Italy that used building materials of volcanic origin. UNSCEAR also noted that in the UK, a value as high as 30 was observed for a house with a high ventilation rate and an unusually low radon concentration (Cliff et al., 1992; UNSCEAR, 2000). The BEIR VI Committee concluded that for dwellings with high radon concentrations, it appears that thoron progeny will not be an important additional source of exposure and dose (NRC, 1999).

(B17) A summary of dose coefficients for thoron progeny, calculated using dosimetric models, is given in Table B.2. Values range from 1.5 to 5.7 mSv per WLM, i.e. 0.42–1.6 mSv per (mJh/m^3) or 10–122 nSv per (Bqh/m^3) of EEC.

(B18) The dose coefficient given in *Publication 50* (ICRP, 1987) is based on the work of an Expert Group of OEC/NEA (1983), which reviewed the models of Jacobi and Eisfed (1980, 1982) and James et al. (1980, 1982). Only doses to the bronchial epithelium and pulmonary tissue were considered.

(B19) In its 1982 report, UNSCEAR not only considered the doses to the lungs based upon the work of Jacobi and Eisfed (1980), but also considered doses to other tissues by applying the dosimetric models given in *Publication 30* (ICRP, 1979). Values of 1.9 mSv per WLM [0.54 mSv per (mJh/m^3)] and 2.5 mSv per WLM [0.71 mSv per (mJh/m^3)] were recommended for indoor and outdoor exposures, respectively. The effective dose coefficients for thoron progeny given in the 1988 UNSCEAR report were based upon the calculations of Jacobi and Eisfeld (1982), and corresponded to an effective dose per unit of potential alpha energy of 0.7 mSv/mJ. These values were retained in the 1993 report (UNSCEAR, 1993) and are given in Table B.2. UNSCEAR (2000, 2009) has since adopted a value of 40 nSv per hBq/ m^3 of EEC [i.e. 1.9 mSv per WLM or 0.54 mSv per (mJh/m^3)] for indoor and outdoor exposures, which is similar to the value given in *Publication 50* (ICRP, 1987).

(B20) The values of the dose coefficients obtained using the HRTM (Marsh and Birchall, 1999; Ishikawa et al., 2007; Kendall and Phipps, 2007) are higher than the values recommended by ICRP (1987) and UNSCEAR (1993). Kendall and Phipps (2007) calculated the effective dose conversion factor for thoron progeny with the HRTM and the most recent biokinetic models for lead (ICRP, 1993) and bismuth (ICRP, 1979). The authors showed that the dose to the lungs typically contributed more than 97% of the effective dose, and that the intake from lead-212 alone represents about 85% of the total dose. Calculations for different age groups (>1

Table B.2. Summary of the calculated dose conversion factors[a] for thoron progeny using direct dosimetry.

Publication	Model type	Exposure scenario	Effective dose[b] (mSv per WLM)	[mSv per (mJh/m^3)]
ICRP (1987)	NEA (1983)	Indoors and outdoors	1.8	0.51
UNSCEAR (1993)	Jacobi and Eisfeld (1982)	Indoors	1.5	0.42
		Outdoors	0.47	0.13
Marsh and Birchall (1999)	HRTM (ICRP, 1994)	Dwellings	3.8	1.1
Porstendörfer (2001)	Zock et al. (1996)	Indoors	2.4	0.68
		Outdoors	2.0	0.56
Ishikawa et al. (2007)	HRTM (ICRP, 1994)	Indoors	5.4	1.5
Kendall and Phipps (2007)	HRTM (ICRP, 1994)	Indoors	5.7	1.6

WLM, working level month; HRTM, Human Respiratory Tract Model.
[a] Calculated for an adult male.
[b] 1 WLM = 4.68×10^4 Bqh/m^3 of equilibrium equivalent concentration of thoron.

year) showed that the dose per unit exposure differed by 10% or less (Kendall and Phipps, 2007).

(B21) Following the decision to treat radon isotopes in the same way as other radionuclides for protection purposes, biokinetic and dosimetric models will be used to provide dose coefficients for radon-220 as well as radon-222.

B.3. References

Baias, P., Hofmann, W., Winkler-Heil, R., Cosma, C., Duliu, O.G., 2010. Lung dosimetry for inhaled radon progeny in smokers. Radiat. Prot. Dosim. 138, 111–118.

Cliff, K.D., Green, B.M.R., Mawle, A., et al., 1992. Thoron daughter concentrations in UK homes. Radiat. Prot. Dosim. 45, 361–366.

Harley, N.H., Cohen, B.S., Robbins, E.S., 1996. The variability in radon decay product bronchial dose. Environ. Int. 22 (Suppl. 1), S959–S964.

Harley, N.H., Pasternack, B.S., 1982. Environmental radon daughter alpha dose factors in a five-lobed human lung. Health Phys. 42, 789–799.

Hofmann, W., Steinhäusler, F., Pohl, E., 1980. Age-, sex-, and weight-dependent dose patterns due to inhaled natural radionuclides. In: Natural Radiation Environment III. CONF-780422, US Department of Energy, 1980, Houston, Texas, USA, pp. 1116–1114.

ICRP, 1979. Limits for intakes of radionuclides by workers, Part 1. ICRP Publication 30. Ann. ICRP 2(3/4).

ICRP, 1987. Lung cancer risk from indoor exposures to radon daughters. ICRP Publication 50. Ann. ICRP 17(1).

ICRP, 1993. Protection against radon-222 at home and at work. ICRP Publication 65. Ann. ICRP 23(2).

ICRP, 1994. Human respiratory tract model for radiological protection. ICRP Publication 66. Ann. ICRP 24(1–3).

Ishikawa, T., Tokonami, S., Nemeth, C., 2007. Calculation of dose conversion factors for thoron decay products. J. Radiol. Prot. 27, 447–456.

Jacobi, W., Eisfeld, K., 1980. Dose to Tissues and Effective Dose Equivalent by Inhalation of Radon-222, Radon-220 and their Short-lived Daughters. GSF-S-626. GSF, Neuherberg.

Jacobi, W., Eisfeld, K., 1982. Internal dosimetry of inhaled radon-222, radon-220 and their short-lived daughters. In: Vohra, K.G., Mishra, U.C., Pillai, K.C., Sadavisan, S. (Eds.), Proceedings of the 2nd

Special Symposium on the Natural Radiation Environment, January 1981, Bombay. Wiley Eastern, New Delhi, pp. 131–143.

James, A.C., Greenhalgh, J.R., Birchall, A., 1980. A dosimetric model for tissues of the human respiratory tract at risk from inhaled radon and thoron daughters. Radiation Protection – a Systematic Approach to Safety, Proceedings of the 5th IRPA Congress, March 1980, Jerusalem, Vol. 2. Pergamon, Oxford, pp. 1045–1048.

James, A.C., Jacobi, W., Steinhäusler, F., 1982. Respiratory tract dosimetry of radon and thoron daughters. The state-of-the-art and implications for epidemiology and radiobiology. In: Gomez, M. (Ed.), Radiation Hazards in Mining: Control, Measurements and Medical Aspects. Soc. Mining Engineers, New York, pp. 42–54.

James, A.C., Birchall, A., Akabani, G., 2004. Comparative dosimetry of BEIR VI revisited. Radiat. Prot. Dosim. 108, 3–26.

Kendall, G.M., Phipps, A.W., 2007. Effective and organ doses from thoron decay products at different ages. J. Radiol. Prot. 27, 427–435.

Kendall, G.M., Smith, T.J., 2005. Doses from radon and its decay products to children. J. Radiol. Prot. 25, 241–256.

Marsh, J.W., Birchall, A., 1999. The thoron issue: monitoring activities, measuring techniques and dose conversion factors. Radiat. Prot. Dosim. 81, 311–312.

Marsh, J.W., Birchall, A., 2000. Sensitivity analysis of the weighted equivalent lung dose per unit exposure from radon progeny. Radiat. Prot. Dosim. 87, 167–178.

Marsh, J.W., Birchall, A., Davis, K., 2005. Comparative dosimetry in homes and mines: estimation of K-factors. Natural Radiation Environment VII. Seventh International Symposium on the Natural Radiation Environment (NRE-VII), May 2002, Rhodes, Greece. Radioactivity in the Environment, Vol. 7. Elsevier Ltd, Amsterdam.

Marsh, J.W., Harrison, J.D., Laurier, D., et al., 2010. Dose conversion factors for radon: recent developments. Health Phys. 99, 511–516.

NCRP, 1984. Evaluation of Occupational and Environmental Exposures to Radon and Radon Daughters in the United States. NCRP Report No. 78. National Council on Radiation Protection and Measurements, Bethesda, MD.

NEA, 1983. Dosimetry Aspects of Exposure to Radon and Thoron Daughters Products. Nuclear Energy Agency Report. NEA/OECD, Paris.

NRC, 1991. Comparative Dosimetry of Radon in Mines and Homes. National Academy Press, Washington, DC.

NRC, 1999. Health Effects of Exposure to Radon. BEIR VI Report. National Academy Press, Washington, DC.

Porstendörfer, J., 2001. Physical parameters and dose factors of the radon and thoron decay products. Radiat. Prot. Dosim. 94, 365–373.

Tokonami, S., Matsuzawa, T., Ishikawa, T., et al., 2003. Changes of indoor aerosol characteristics and their associated variation on the dose conversion factor due to radon progeny inhalation. Radiosotopes 52, 285–292.

UNSCEAR, 1982. Sources and Effects of Ionizing Radiation. 1982 Report to the General Assembly with Annexes. United Nations, New York.

UNSCEAR, 1988. Sources, Effects and Risks of Ionizing Radiation. 1988 Report to the General Assembly with Annexes. United Nations, New York.

UNSCEAR, 1993. Sources and Effects of Ionizing Radiation. 1993 Report to the General Assembly with Scientific Annexes. United Nations, New York.

UNSCEAR, 2000. Sources and Effects of Ionizing Radiation. 2000 Report to the General Assembly with Scientific Annexes. United Nations, New York.

UNSCEAR, 2009. UNSCEAR 2006 Report, Annex E. Sources-to-Effects Assessment for Radon in Homes and Workplaces. United Nations, New York.

Winkler-Heil, R., Hofmann, W., 2002. Comparison of modelling concepts for radon progeny lung dosimetry. Proceedings of Fifth International Conference – High Levels of Natural Radiation and

Radon Areas: Radiation Dose and Health Effects, September 2000, Munich. Elsevier, Amsterdam, pp. 169–177.

Winkler-Heil, R., Hofmann, W., Marsh, J.W., Birchall, A., 2007. Comparison of radon lung dosimetry models for the estimation of dose uncertainties. Radiat. Prot. Dosim. 127, 27–30.

Zock, C., Porstendörfer, J., Reineking, A., 1996. The influence of the biological and aerosol parameters of inhaled short-lived radon decay products on human lung dose. Radiat. Prot. Dosim. 63, 197–206.

Statement on Radon

ICRP PUBLICATION 115, PART 2

Statement on Radon

ICRP PUBLICATION 115, PART 2

Approved by the Commission in November 2009

(1) The Commission issued revised recommendations for a system of radiological protection in 2007 (ICRP, 2007), which formally replaced the Commission's 1990 Recommendations (ICRP, 1991) and updated, consolidated, and developed the additional guidance on the control of exposure from radiation sources. The Commission has previously issued recommendations for protection against radon-222 at home and at work in *Publication 65* (ICRP, 1993).

(2) The Commission has now reviewed recently available scientific information on the health effects attributable to exposure to radon and its decay products. The Commission's full review accompanies this Statement. As a result of this review, for radiological protection purposes, the Commission now recommends a detriment-adjusted nominal risk coefficient for a population of all ages of 8×10^{-10} per (Bqh/m^3) for exposure to radon-222 gas in equilibrium with its progeny (i.e. 5×10^{-4} per WLM). The Commission's findings are consistent with other comprehensive estimates, including that submitted to the United Nations General Assembly by the United Nations Scientific Committee on the Effects of Atomic Radiation (UNSCEAR, 2009).

(3) Following the 2007 Recommendations, the Commission will publish revised dose coefficients for the inhalation and ingestion of radionuclides. The Commission now proposes that the same approach be applied to intakes of radon and its progeny as that applied to other radionuclides, using reference biokinetic and dosimetric models. Dose coefficients will be given for different reference conditions of domestic and occupational exposure, taking into account factors including inhaled aerosol characteristics and disequilibrium between radon and its progeny. Sufficient information will be given to allow specific calculations to be performed in a range of situations. Dose coefficients for radon and its progeny will replace the *Publication 65* dose conversion convention which is based on nominal values of radiation detriment derived from epidemiological studies comparing risks from radon and external radiation. The current dose conversion values may continue to be used until dose coefficients are available. The Commission advises that the change is likely to result in an increase in effective dose per unit exposure of around a factor of two.

(4) The Commission re-affirms that radon exposure in dwellings due to unmodified concentrations of radium-226 in the earth's crust, or from past practices not conducted within the Commission's system of protection, is an existing exposure situation. Furthermore, the Commission's protection policy for these situations continues to be based on setting a level of annual dose of around 10 mSv from radon where action would almost certainly be warranted to reduce exposure. Taking account of the new findings, the Commission has therefore revised the upper value for the ref-

erence level for radon gas in dwellings from the value in the 2007 Recommendations (ICRP, 2007) of 600 Bq/m^3 to 300 Bq/m^3. National authorities should consider setting lower reference levels according to local circumstances. All reasonable efforts should be made, using the principle of optimisation of protection, to reduce radon exposures to below the national reference level. It is noted that the World Health Organization now recommends a similar approach (WHO, 2009).

(5) Taking account of differences in the length of time spent in homes and workplaces of about a factor of three, a level of radon gas of around 1000 Bq/m^3 defines the entry point for applying occupational protection requirements for existing exposure situations. In *Publication 103*, the Commission considered that the internationally established value of 1000 Bq/m^3 might be used globally in the interest of international harmonisation of occupational safety standards (ICRP, 2007). The Commission now recommends 1000 Bq/m^3 as the entry point for applying occupational radiological protection requirements in existing exposure situations. The situation will then be managed as a planned exposure situation.

(6) The Commission re-affirms its policy that, for planned exposure situations, any workers' exposure to radon incurred as a result of their work, however small, shall be considered as occupational exposure (see Para. 178 of ICRP, 2007).

References

ICRP, 1991. 1990 Recommendations of the International Commission on Radiological Protection. ICRP Publication 60. Ann. ICRP 21 (1–3).

ICRP, 1993. Protection against radon-222 at home and at work. ICRP Publication 65. Ann. ICRP 23 (2).

ICRP, 2007. The 2007 Recommendations of the International Commission on Radiological Protection. ICRP Publication 103. Ann. ICRP 37 (2–4).

UNSCEAR, 2009. UNSCEAR 2006 Report. Annex E. Sources-to-Effects Assessment for Radon in Homes and Workplaces. United Nations, New York.

WHO, 2009. WHO Handbook on Indoor Radon: a Public Health Perspective. WHO Press, Geneva.

Annals of the ICRP

Published on behalf of the International Commission on Radiological Protection

Aims and Scope

The International Commission on Radiological Protection (ICRP) is the primary body in protection against ionising radiation. ICRP is a registered charity and is thus an independent non-governmental organisation created by the 1928 International Congress of Radiology to advance for the public benefit the science of radiological protection. The ICRP provides recommendations and guidance on protection against the risks associated with ionising radiation, from artificial sources widely used in medicine, general industry and nuclear enterprises, and from naturally occurring sources. These reports and recommendations are published approximately four times each year on behalf of the ICRP as the journal *Annals of the ICRP*. Each issue provides in-depth coverage of a specific subject area.

Subscribers to the journal receive each new report as soon as it appears so that they are kept up to date on the latest developments in this important field. While many subscribers prefer to acquire a complete set of ICRP reports and recommendations, single issues of the journal are also available separately for those individuals and organizations needing a single report covering their own field of interest. Please order through your bookseller, subscription agent, or direct from the publisher.

ICRP is composed of a Main Commission, a Scientific Secretariat, and five standing Committees on: radiation effects, doses from radiation exposure, protection in medicine, the application of ICRP recommendations, and protection of the environment. The Main Commission consists of a Chair and twelve other members. Committees typically comprise 10–15 members. Biologists and medical doctors dominate the current membership; physicists are also well represented.

ICRP uses Working Parties to develop ideas and Task Groups to prepare its reports. A Task Group is usually chaired by an ICRP Committee member and usually contains a number of specialists from outside ICRP. Thus, ICRP is an independent international network of specialists in various fields of radiological protection. At any one time, about one hundred eminent scientists and policy makers are actively involved in the work of ICRP. The Task Groups are assigned the responsibility for drafting documents on various subjects, which are reviewed and finally approved by the Main Commission. These documents are then published as the *Annals of the ICRP*.

International Commission on Radiological Protection